Photovoltaics Guidebook
for Decision-Makers

Springer-Verlag Berlin Heidelberg GmbH

Physics and Astronomy ONLINE LIBRARY

http://www.springer.de/phys/

Achim Bubenzer
Joachim Luther (Eds.)

Photovoltaics Guidebook for Decision-Makers

Technological Status and Potential Role in Energy Economy

With 85 Figures, 39 Tables

Springer

Prof. Achim Bubenzer
Fachhochschule Ulm
89075 Ulm, Germany
e-mail: bubenzer@fh-ulm.de

Prof. Joachim Luther
Fraunhofer Institut
für Solare Energiesysteme
Oltmannstr. 5
79110 Freiburg, Germany
e-mail: luther@ise.fhg.de

The cover picture is based on Figure 1.1 of this book.

ISBN 978-3-642-07457-8

Library of Congress Cataloging-in-Publication Data

Photovoltaics guidebook for decision makers: technological status and potential role in energy economy/Achim Bubenzer, Joachim Luther (eds.).
p. cm. Includes bibliographical references and index.
ISBN 978-3-642-07457-8 ISBN 978-3-662-05140-5 (eBook)
DOI 10.1007/978-3-662-05140-5
1. Photovoltaic power systems. I. Bubenzer, Achim. II. Luther, J. (Joachim)
TK 1087.P48523 2002
333.792'3–dc21 2002026900

http://www.springer.de

Originally published by Springer-Verlag Berlin Heidelberg New York in 2003
Softcover reprint of the hardcover 1st edition 2003

Typesetting: Data prepared by the author using a Springer TEX macro package
Final page layout by EDV-Beratung Frank Herweg, Leutershausen
Cover design: *design & production* GmbH, Heidelberg

Printed on acid-free paper SPIN 10751344 57/3141/tr 5 4 3 2 1 0

Preface

Mankind has to introduce sustainable development within the 21st century if basic life conditions are not to be put at risk. Above all, this demands eventually taking note of the limitations of non-renewable energy resources and of the limited capacity of local and global ecosystems to deal with disposal and emissions. For the resulting problems, we have to develop solutions. But sustainability is more than the protection of the natural environment. Justice, social stability, and adequate opportunities for education and employment for all people are of similar importance for our future. Both a transformation of energy policy in industrialized countries and a new policy for support of the less developed countries are necessary to prepare the way for a sustainable global community.

Today we are far from such sustainable development. Global energy consumption and the resulting emissions are still rising considerably and increasingly endanger the ecological stability of the earth. Growing scarcity and rising prices of fossil fuels may lead to economic and political instability in many countries in the near future. In addition to this critical development there exists an extreme divide in prosperity between industrialized and less developed countries, which is partly caused by an imbalance in access to sufficient and affordable energy, and which itself leads to conflicts and environmental damage.

The increasing use of renewable energy sources can contribute significantly to solving these problems. These resources are able to meet global energy demand many times over and can therefore remove or at least diminish numerous causes of the above-mentioned dangers and risks. The extensive use of renewable energies worldwide would be a very important step towards sustainability.

The great potential of renewable energies to solve the quantitative energy problems of the world without damaging it has been recognized in energy policy, at least in principle, in recent years.

For instance, the European Union intends to increase the contribution of renewable energies to 12 % by the year 2010. This is also on the agenda in Germany which aims to double the share of renewable energies by 2010. Growing pressure to fulfill international obligations of climate protection is mainly responsible for the formulation of these goals. In Germany the planned removal of nuclear power plants will give additional momentum to changes in energy policy. Chances have never been better to give energy policy a new orientation.

But simultaneously with the formulation of these ambitious goals to use renewable energies more intensively, the basic conditions in energy policy and economy have changed considerably. The electricity market is in an irrevocable process of liberalization, and national markets integrate into international market networks. In addition to the oil companies already acting globally, electricity and gas utilities are joining together and withdrawing themselves more and more from national standards. National policies orientated towards long-term welfare and economic effectiveness of society as a whole are confronted with decisions of energy companies aiming at short-term economic profit.

But radical change can also lead to new chances and open new options. Energy utilities working until recently only in monopolies now have to advertise for clients and must offer attractive energy services. New participants gain opportunities to have a share in the electricity market. Businesses and citizens can change their electricity supplier and are able to influence the way electricity is generated. Energy supply structures are already in a process of reconstruction. Decentralized solutions for supplying energy are gaining ground. This trend is supported by information and communication technologies, which enable control and supervision of many small and interconnected energy generation units. Liberalization of the energy market accelerates this tendency, because it demands short-term financing, planning and construction of plants, and ready adaptability to steadily changing conditions.

On the whole, there exist good prerequisites for increased use of all renewable energy technologies. Photovoltaics, as one of the most important long-term technologies for using solar energy, will strongly profit from these developments. Therefore it is of particular interest to study in more detail the potentials and abilities of photovoltaics in the restructuring of our energy supply system.

In this book we try to approach these topics from different viewpoints. Based on the global challenges for the future energy supply, in the first chapter we will derive the resulting demands which photovoltaics has to meet if this technology is to play a relevant role (in our energy system) within the coming decades. The global and national technical potential of photovoltaics is illustrated, and future market developments necessary to meet the anticipated shares in energy contribution are outlined. In this way we gain important information on the course and the intensity in which the development of photovoltaics should continue.

In addition it should be kept in mind that although photovoltaics is assessed here from an energy economy point of view, the present motivation for financial investments in photovoltaic technology is mostly a different one: to a large extent photovoltaics today satisfies the customer-specific need for an off-grid power supply for small- and medium-size electrical systems. The ability to serve this type of need, as well as large scale energy economy demands, is a unique feature of photovoltaic technology. No other renewable energy technology is able to cover applications with a bandwidth of more than eight orders of magnitude of electrical power: from a 10 mW-system (1 cm^2) to a several MW-system (several 10 000 m^2). This unique variability enables photovoltaic technology to build capacities and markets gradually and thereby reach a production volume sufficient to serve relevant energy demands in the long term.

In the second chapter we provide a technical overview of the technological status and potential of photovoltaic modules and systems. This overview is meant for the technically interested reader who is not necessarily technically trained or a photovoltaics expert.

The important topic of industrial production technology and cost reduction for photovoltaic systems is discussed in Chap. 3. For this topic we tried to include current inputs from several important industrial players in the field of production of photovoltaic system components.

The fourth chapter addresses some deeply-rooted prejudices concerning ecological and life cycle assessment, for example the widespread confusion and uncertainty about energy payback time of photovoltaic systems.

Finally we think it is important to review and discuss public as well as industrial strategies for R&D and market introduction which are currently practiced or which have been applied in the past. This is done in Chap. 5.

The reader short of time will find the essentials of each chapter in the corresponding section of the executive summary at the beginning of this book. Except for Chap. 2, which describes the technological status, the topics dealt within this book are quite complicated subjects which often give rise to controversial discussions. Therefore the summaries of the following chapters are structured in the same way: First of all the authors state what they consider to be clear and solidly based facts ("What's clear"). Secondly the aspects which will need further clarification are listed ("What's still an open question"), and finally the authors describe actions needed to solve problems related to their specific subject ("What needs to be done").

The goal of writing a book which comprises all relevant aspects required to evaluate photovoltaic technology as an industrial and energy-economy option, and which could indeed qualify as "Photovoltaic Guide for Decision Makers", needed technical as well as economical competence from a sufficiently large group of experts. This is why the group of authors started the project of this book with a workshop, which took place in October 1999 at the Reisensburg castle near the city of Ulm in Germany. Invited participants in this workshop were experts in the fields of photovoltaic module and system technology, energy economy, ecology, business, public administration, political economy, and public utilities (see list on p. 259). The workshop was meant to stimulate discussion about the topics of this book, and to initiate cooperation among participants and authors to answer open questions. Participants were selected in such a way that there was a representative range of different points of view, which enabled constructive disputes.

The authors of the separate chapters however bear total responsibility for the contents of this book. This means that some aspects or conclusions stated in this book may not be shared by all participants of the Reisenburg workshop.

The whole project of workshop and book was organized by the Ulmer Initiativkreis nachhaltige Wirtschaftsentwicklung (unw) e.V. It was made possible by generous support from the Solarstiftung Ulm/Neu-Ulm and the Stiftung Landesgirokasse, as well as by donations from Bayernwerk AG, Würth Solar GmbH, SMA-Regelsysteme, and Steca GmbH. The idea for the project grew out of a fruitful cooperation between the Fachhochschule Ulm (Ulm University of Applied Sciences) and the Zentrum für Sonnenenergie- und Wasserstoffforschung in Ulm, Germany. We thank both institutions for their friendly support and assistance.

We thank G. Lindemann for expert help in computerized typesetting with LaTeX and compiling the separate contributions to this book.

Ulm, Freiburg, *A. Bubenzer*
August 2002 *J. Luther*

Contents

Executive Summary

Strategies for Introducing Renewable Energies and the Contribution of Photovoltaics

J. Nitsch and F. Staiß

What's Clear

Global and national scenarios of primary and in particular electrical energy consumption for the coming decades basically all predict a strong increase in technical utilisation of renewable energies (RES). Only a significant increase in the use of RES satisfies the requirements of climate protection and allows suitable growth in energy consumption for newly developed and developing countries. At the same time, this aims at decreasing the share of fossil primary energies in the medium and long term, and has the potential to reduce the share of nuclear energy to zero. To achieve this goal, however, all RES technologies (hydro power, wind power, biomass, geothermal heat, and solar irradiation) have to be mobilized in a balanced way and in coordinated time sequences, i.e. according to their economic market relevance and their technical potential. In this case their necessary dissemination will not be limited by todays low economical potential.

Among the RES, solar irradiation will in the long term have to become the main contributor to a global renewable energy supply because of its virtually unlimited potential. Therefore both technologies of using solar irradiation for electricity supply, i.e. solar thermal and photovoltaic electricity generation, will be essential for a future sustainable global energy supply, although their current contribution is still negligible. The two technologies complement one another in a very favorable way. Points which favor solar thermal electricity generation are low costs as well as the possibilities of combining it with existing power plants by means of hybrid operation, and incorporating it into grid structures. Photovoltaics is favored by its enormous flexibility with respect to size and fields of application, by its existing competitive partial market, i.e. especially off-grid electricity supply, and its robustness, its long life-time and low required maintenance. In the longer term, the remarkable potentials for cost reduction due to increasing markets and technological progress will be of growing importance. The dissemination of both technologies will not be limited by low technical potentials at all. In particular, the total technical potential of photovoltaic electricity generation is about four times present global electricity consumption. Assuming optimistic growth rates, only 5 % of this global technical potential will be used by 2050.

In Order to enable photovoltaics to play a part in shaping the future energy supply in an economically and energetically relevant way in the next few decades, its contribution to global electricity generation will have to reach at least 0.1 % in 2010 and 1 % in 2020. With a basis like that, the share of photovoltaics will probably achieve a good 10 % in 2050. Thus, a balanced, sound share of photovoltaics could be reached in the total RES contribution, which could then be approximately 70 % according to the scenarios. It is important to note that in 2050 the global demand for electricity will at least have about doubled as compared to present consumption.

However, the key point in assessing the potential of global photovoltaic electricity generation is the following: in order to reach the shares for photovoltaics mentioned above, average growth rates for the annual global installations of PV systems of about 30 %/yr are necessary until about 2010. Later on they may level off, from 2020 on decreasing to 10 %/yr and below. The recent market growth is indeed consistent with this requirement: the market for photovoltaic systems has undergone very dynamic development within the last two decades. In the 1990s production quadrupled, which corresponds to an average growth rate of 17 %/yr. In the last three years growth rates of more than 30 %/yr have been reached. Since the early 1980s, prices (not production costs) for PV modules have fallen by a factor of 3. The international PV industry has reached an annual turnover of approximately 1.5 billion US$, one third of which falls to PV modules. It is worthwhile to note that – in contrast to other renewable energy technologies – PV module production is in the hands of eight big industrial companies accounting for 80 % of the world market. These top eight can afford the high investments in production facilities and development and can rely on a global distribution network. On the other hand, it should be noted that the remaining two thirds of the PV system turnover – the part excluding the modules – is accessible to smaller companies as well.

At present (2002), the typically highly subsidized market of grid-connected systems amounts to about 40 % of the global PV market. It has grown considerably in the last five years. The market for commercial applications not receiving any public subsidies, however, amounts to 50 to 60 % of the total world market. This market comprises decentralized applications with lower electricity demand, such as the so-called consumer market (pocket calculators, watches), the leisure market (camping, garden), professional applications (cathodic protection, military applications, signaling equipment), solar home systems, water pumping systems and village electrification systems. This means that photovoltaics has a sound commercial basis on which part of the necessary market growth can be developed.

What's Still an Open Question

The key question which remains open is whether the necessary market growth in photovoltaic systems will be reached during the next decade. The answer to this question largely depends on political will in the public sector and on the ability and will for long-term strategic planning in industry.

What Remains To Be Done

If the annual market growth is less than 20 % in the near future, a share of more than 1 % in electricity supply will probably be reached between 2030 and 2040 at the earliest. This would not be enough to influence future energy supply significantly and in time. It is therefore of prime importance to stabilize the emerging acceleration of market growth in the next decade or expand it even further. For this purpose, effective instruments for market support like for example in Germany the Act on Granting Priority to Renewable Energy Sources in the electricity grid are crucial preconditions. They are required in similar form by all the industrialized countries to quickly increase the big market segment of grid-connected power plants and to enable considerable cost reductions. With the industrialized countries acting as pioneers, the further dissemination in the field of off-grid supply in countries with less developed infrastructure would then be decisively facilitated.

Solar Modules and Photovoltaic Systems

J. Luther, K. Preiser, and G.Willeke

Solar cells are electronic devices that convert sunlight directly into electricity. They are composed of semiconducting material, which has electrical properties somewhere between those of a metal like copper and an insulator like glass. Important semiconductors that are currently being studied for solar cell applications are crystalline silicon, $CuInSe_2$ (CIS), CdTe, GaAs, and amorphous silicon hydrogen alloys. In order to achieve good conversion efficiency of light into electricity, a high structural and electronic quality semiconductor layer is usually required, which must be free of defects like unwanted impurities and crystal imperfections to an extremely high degree. Surfaces which represent areas of crystal imperfections have to be well passivated by depositing high quality covering layers. A charge separation structure has to be formed in the semiconductor, usually by incorporating well-controlled small amounts of wanted impurities, so-called dopants, or by depositing layers with suitable properties. Light that enters the solar cell excites positive and negative charge carriers throughout the device which have to be transported to and collected at the surface. Nowadays good solar cells have a conversion efficiency in the range 10–20 %.

Fabricating solar cells on a large industrial scale requires simple automated processes with high machine availability and good manufacturing yield that result in stable, high solar cell efficiencies. Crystalline silicon (Si) wafer solar cell technology (market share 90 % with increasing tendency) apparently fulfills these requirements. It benefits from the fact that the high-rate and large-volume crystal growth and wafer preparation is decoupled from the actual cell manufacturing, and that both front and back surfaces of the wafer are available for further processing, leading to the resulting high solar cell efficiencies. In addition, the sorting and selection of individual cells for optimum current and power matching in solar modules is another positive feature that leads to the comparatively simple and defect-tolerant high efficiency processes required in mass production. A large number of manufacturing lines with an annual production capacity of roughly 10 MW have each been operating for many years; in the year 2000 they produced 260 MW of crystalline silicon solar cells. Despite decades of industrial production experience with this technology, fully automated continuous in-line manufacturing facilities are just being introduced. Key issues in processing are wafer handling, which is of increasing importance due to the cost-driven decreasing wafer thicknesses, and the implementation of cost-effective efficiency-enhancing technologies in the production environment.

A radically different approach from the separate handling of self-supporting crystalline silicon solar cells is taken in thin film technologies like amorphous Si, CIS, and CdTe. Here, very thin layers a few micrometers thick – a fraction of the thickness of a human hair – are deposited from the gas phase onto a suitable foreign low-cost substrate. In the most successful technologies these depositions take place at such low temperatures that window glass can be used. However, these low film growth temperatures lead to small-grained polycrystalline or even amorphous layers which contain many structural and electronic defects, and are therefore of lower electronic quality than crystalline silicon wafers. The comparatively small film growth rates obtainable lead to the necessity for scaling up the deposition processes to m^2 sizes in order to achieve significant production volumes. If this large area were manufactured into a single solar cell, it would lead to impractically large currents. Therefore the whole area has to be subdivided into several tens of individual cells, which are then series-interconnected in an integrated process. This leads to the fact that process homogeneity is extremely stringent, since the cell with the smallest current limits the output of the whole module, and the removal of malfunctioning cells is in general not possible. Alltogether, the requirements of fast, homogeneous, and high electronic quality depositions on large areas that lead to stable module efficiencies high enough for low-cost applications are extremely demanding, and are an important reason why thin-film technologies so far play no major role in photovoltaic power generation.

Among the thin film materials a-Si is the most extensively researched, and despite a 20-year production history there is no 10 MW-line yet that operates at full capacity. The total market share for 2000 of almost 10 % (= 27 MW) is almost entirely used in low-efficiency consumer applications. The two major hurdles that need to be overcome are the limited efficiency and the photodegradation intrinsic to the material. This problem is, at least partially, eliminated at the cost of a much more complex technology by depositing two or three solar cells on top of one another.

The thin-film candidates CdTe and CIS (and its predecessor CuS/CdS, whose development was terminated in the beginning of the 1980s because of stability problems) have been studied over the last 40 years but are still in the pilot-type production stage. It remains to be seen whether the good efficiencies obtained on small-size laboratory cells can be reproduced in an industrial scale production under the conditions of high throughput and high yield.

If the market for crystalline silicon solar cells continues growing at the present rate, a serious silicon feedstock shortage will be felt at the latest by the year 2005. Until then a sustainable source independent of the microelectronics industry of low-cost so-called solar grade silicon has to be established. The supply of the raw material low grade silicon is practically unlimited. The Te and In resources, essential for CdTe and CIS cells, do not limit the deployment of these solar cell technologies in the next 20 years.

GaAs concentrator cells, thin-film crystalline silicon solar cells, as well as dye-sensitized and organic solar cells, are still in the early stages of research and their capabilities need to be further assessed.

In order to protect solar cells from the environment, they have to be encapsulated in modules. Typical module sizes range from 10 W to several 100 W. For crystalline Si modules operational lifetimes of up to 25 years are warranted. In professional applications small solar cells may be integrated into the surface of industrial products.

The modular nature of photovoltaic generators – consisting of photovoltaic modules – means that energy supply systems can be constructed for an extremely wide power range. The power spectrum ranges from a few milliwatts for watches or calculators, to kilowatt systems as remote power supplies, e.g. for mountaineering lodges or water pumps, up to large central photovoltaic power stations in the megawatt range.

Photovoltaics is generally considered to be an expensive method of supplying electricity. On the other hand, there is a growing non subsidized market for this kind of energy technology all over the world. This indicates that solar electricity has entered areas where its application is already economically viable. In parallel with cost reductions, these areas will grow and extend to other fields of electricity supply.

A main market segment is the power supply of appliances and other small loads. Applications are either grid-independent or remote from the grid. In these cases the reason for applying photovoltaics is generally its cost effectiveness. Other advantages besides the cost effective production of electricity are reliability, portability, environmentally benign production of energy, etc. In many cases these advantages exist even if the consumer is situated near a grid. While the majority of PV modules has been sold to power these off-grid systems, in 1999 already 40 % of all modules have been used in systems connected to the public electricity grid. These applications include both smaller, decentralised, grid-connected systems, with the solar modules typically installed on house roofs, as well as large, central, grid-connected systems, with large areas of solar modules usually mounted on dedicated support structures.

Rural electrification with photovoltaics has become a rapidly emerging market in the last few years. While large area grid extension is getting more and more difficult due to the high costs of connecting remote households, with their relatively low energy consumption for lighting and communication, about two billion people worldwide remain without access to grid electricity. As the population grows rapidly, especially in rural areas of developing countries, there are more people each day without access to electricity than the day before. This is the main reason why in many countries rural electrification programs have been initiated. The World Bank, as one of the major support organisations in rural electrification, is involved in more or less all these national programs, thus helping to cover the high up-front costs of PV rural electrification and establishing innovative financing schemes. Most of these programs today concentrate on Solar Home Systems, as these systems are relatively easy to install and operate. The key to success or failure of these programs lies in the establishment of appropriate infrastructures for installation, operation, and maintenance of these systems, as well as in the introduction of schemes for quality assurance, microfinancing, and adequate involvement of the users. In the next future photovoltaic village power systems may complement Solar Home Systems in the area of rural electrification.

In all standalone applications, the storage unit is one of the technical components which still needs the most development. Often described as the weakest link in the chain, it is responsible for a large share of the costs in long-term operation. Thus optimization of the interaction of all components in the system with the aim of reducing the overall lifetime costs is today the main issue on this sector. As the battery industry has also recognized, there is a great need for the development of better battery management principles, adaptation of the existing battery technologies to the charging and discharging requirements in photovoltaic systems, and the development of completely new and better suited storage units.

Grid-connected photovoltaic systems are being introduced in large quantities in several national programs, like the 70 000-roofs Program in Japan and the 100 000-roofs Program in Germany. Besides small rooftop systems, more and more large central power stations are being erected by utilities or industries on facades, roofs, or in open fields. Besides the pure aspect of electricity production, grid-connected photovoltaic systems do have an enormous effect on, e.g. public relations for utilities and enterprises. Therefore quality control, monitoring, and dissemination of the results gained with these installations are key factors for the success of these installations and their further dissemination.

As a result of the liberalization of the European electricity market and electricity distribution, grids are subject to significant technical and organizational changes. The process has just started and will continue for years. Major results of the liberalization are the organizational separation of power generation, transmission, and distribution companies along with supply companies and the open market for all participants, resulting in fierce market competition and price decline. This again will lead to an increased number of distributed smaller and medium-sized power generators, an increasing replacement of traditional means of power generation by a mix of different technologies (like combined heat and power units, renewable energies) and partly increasing demands of customers on electricity quality and availability. In the case of grid extension and grid upgrading photovoltaic may show, due to its modularity, considerable economic advantages. Especially for peak load production (e.g. to power air conditioning equipments) photovoltaics seems to be an appropriate technical solution.

Grid-connected photovoltaic systems typically do not need battery or other storage units. A strong bidirectional electrical grid is able to handle stochastic short term fluctuations due to moving clouds, medium term trends caused by weather situations, as well as seasonal variations. From simulation calculations it can for example be concluded that under the assumption of a strong bidirectional grid and stringent load management measures it is feasable to feed at least 35 % of highly fluctuating electricity from PV and wind sources into the German electricity grid. From a strategic point of view these findings suggest that in parallel with the installation of of PV electricity generation capacity, the grid structure, the system of complementary electricity production , and the load management technologies must be adapted to the new energy supply scheme.

Industrial Strategy and Cost Reduction Assessment

T. Schott

What's Clear

It is a general consensus that the industrial level of photovoltaic production achieved so far guarantees a further market expansion by at least a factor of ten until the year 2010. This prognosis includes continued technological learning, leading to a halving of the system price from the present values. Despite this expected progress, however, another decade of PV development will be required to bring the cost of PV-generated electricity to the economic level appropriate for large scale application in the energy economy of fully industrialized countries.

Such an ambitious scenario demands steady technological progress, combined with a similarly continuous upscaling of manufacturing capacities. Annual growth rates around 20 to 30 % in the market provide a speed of development optimal in the sense that both production experience and basic innovations have enough time to be absorbed in the engineering and investment cycles of the expanding PV economy. This scenario is now agreed upon by a majority of managers in the PV industry.

In the years to come, new PV manufacturing plants will approach production capacities of 100 MW. Such a plant will contain several production lines, typically providing a 10 MW production capacity. The solar cell production will still be dominated by silicon wafer-based technologies. A growing number of advanced thin-film pilot productions will gradually enter the market. Both groups of technologies will strongly compete for the same applications. However, thin-film technologies shall also enable new applications and markets. Both groups of technologies will follow a route of technological development, leading to the anticipated price reduction of 50 % according to the learning curve scenario.

It is now generally believed that cost reduction during the subsequent decade until 2020 can be linearly extrapolated along the learning curve. However, the speed of progress is hard to forecast and will be extremely dependent on the boundary conditions. This uncertainty is due to the fact that progress will be dominated by investment decisions on an industrial scale which, in turn, depend on the expected market demand and on the standardisation of applications. Innovations, on the other hand, will be pushed by this market growth and implemented in the production under favourable conditions. The recent market expansion of wind power in Germany during the passed decade may provide an example of what is possible within a relatively short period, having reached a share of electricity supply in the range of a few percent. It is believed that under such favorable conditions, another factor of two may be achieved for the cost reduction.

Beyond the year 2020, the development is more obscure and the stability of the learning curve is less certain. Further cost reduction by a factor of two to three requires new technologies which are at best visible at the laboratory stage today. Such new horizons (like organic cells, flexible cell structures, largely integrated applications etc.) are well within the scope of current research and pre-industrial approved goals. It is a general conviction that a great effort in research and development will be necessary to reach this goal.

What's Still an Open Question

The PV learning curve and the linear extrapolation of the double logarithmic representation are assumed as the basic guide line of development. This progress is suggested by an impressive number of examples in the history of technological progress. As a "best guess", the learning curve theory allows coupling the innovative process with more non-technical boundary conditions like investment cost, market development and market introduction subsidies. However, even if this scenario may seem very convincing, there is no guarantee that this will also happen in the case of PV. It remains an open question whether such "causality" in development can be assumed.

Concerning the non-technical boundary conditions, it remains a question whether the energy economy and the governments will accept such a long-range goal derived from the learning curve extrapolation. Renewable energies are even considered by many decision makers as a minor energy option not of true strategic importance. It is unknown whether the growing number of investors and politicians accepting the transition to new forms of energy supply will awaken the appropriate political support within due time.

Even if the necessary political consent and long-term commitment can be achieved, another question remains open: namely the speed of investment. PV is one of the most capital-intensive technologies in the energy sector (perhaps comparable to nuclear or hydroelectric power), and the cost of power production strongly depends on purely economic parameters like interest rates, depreciation periods and so on. Historically, such investments have been decided upon for reasons of general economic development, oriented on a long-term scale, and taken as a precaution for future resources and independence (like nuclear energy or large hydroelectric power plants). In such cases, the risk of investment usually has been taken by the public. It remains an open question, whether society is now ready to consent to environmental policy beyond the short- and medium-term conditions.

What Remains To Be Done

As the speed of progress along the PV learning curve depends on technological innovation as well as on political boundary conditions, a reasonable strategy has to involve both. For a PV policy, more than for any other renewable energy option, the weight of environmental benefits (in relation to the irreversible damages caused by the present energy supply system with its high external cost) has to be accounted for in the cost balance. Any political commitment over such long periods, as needed in the case of PV, must be carried by a general conviction, releasing the necessary forces and resources in society and economy. With respect to this political uncertainty, the large technical potential of PV and its proven feasiblity are well within the range of economic and industrial capability. The special role of PV as the renewable energy source option with the highest technical and environmental potential, but also with the highest economical risk, must be accepted first.

The final benefit of PV technology promotion in the form of future employment and leading in exports can only be realised with an increased and early commitment to R&D to provide the "next generation" technologies. This historical route has been followed for all key technologies we know and remains a general challenge long after such technologies have realised profitable markets. Industry and research must be guided to appropriate structures which are adapted to certain stages of development, leaving the early stage of primary technology orientation for a primary market-oriented approach. For the next fifty years (a time span not too long in terms of an energy system), the fundamental research for future options must be intensified.

It can be concluded that the market pull generated by subsidies (like the 100 000 roof program or the feed-in law in Germany) has been sufficient to promote the domestic industrial technology and the market. To conclude that the industry should now take over or self-finance further R&D efforts, however, would be misguided. The technological basis is still fragil and dominated by near-term development. An ongoing dialogue between policy, industry and research is necessary to analyse the appropriate boundary conditions which encourage the best technological learning rate with the most efficient and economic learning investment. This dialogue is also overdue concerning the already global competition of PV industries.

As soon as such a political long-term strategy is established, the public "learning investment" is well within acceptable levels, and the economic risk is still manageable. Any delay, for instance a "stop and wait for higher energy prices", will make PV introduction and development more expensive in the long run. The long tradition of PV R&D in Germany, with extraordinary scientific and technological results, makes it fertile to continue the described development.

Life Cycle Assessment of Photovoltaic Systems – Trying to Fight Deep-Seated Prejudices

M. Pehnt, A. Bubenzer, and A. Räuber

What's Clear

Although a detailed and complete life cycle assessment (LCA), describing the environmental impact of a modern grid-connected photovoltaic system, is not available today, some basic and reliable statements can be made.

PV systems on the basis of today's "workhorse" in PV technology, namely crystalline Si modules, do not show any relevant weak points in an LCA such as danger to human health and safety, impact on ecosystems, or natural resource depletion. Some confusion with respect to LCA was generated by assessing modules made from off-spec monocrystalline silicon supplied by the semiconductor industry. This material, which is processed twice, has an extremely high energy demand. However, the supply of off-spec electronic grade silicon will in a few years no longer be sufficient to cover the demand of the PV industry. Therefore this technological path will play no major role for the PV industry in the future.

The life cycle CO_2 emissions are mainly due to the electricity demand during the production of PV modules. However, the specific CO_2 emissions which can be attributed to the electrical energy produced by grid-connected PV systems using polycrystalline modules made from solar grade silicon are only 20 % of the specific CO_2 emissions from conventional electricity generation with an average power plant mix. These remaining CO_2 emissions, however, are a burden imported from fossil energy consumption during production of PV modules and are thus not inherent to PV technology. In addition, other PV technologies, especially thin-film cells, promise a further reduction in materials and process energy demand. For these technologies, however, the data quality is much lower due the small production experience gained so far.

An aspect of LCA which has been discussed extensively in the past is the energy payback time (EPBT) of PV systems. Some of the confusion about EPBT was due to an unclear definition of this term. However there is an unambiguous definition of EPBT accepted and used in energy economy (see Sect. 4.2.1). EPBTs of grid-connected PV systems based on polycrystalline solar grade Si are two years in southern and three years in middle Europe. Since the EPBT does not contain any information on the remaining lifetime of the system, we defined a complementary term: Net Energy Life Time (NELT). NELT is defined as lifetime of the system minus EPBT. NELT is therefore the time during which a grid-connected PV system will supply electrical energy free of any energy-related emissions. Assuming a lifetime for a PV system of 25 years, NELT will be more than 20 years for any grid-connected PV system. Nevertheless, these EPBT or NELT data leave much room for improvement, since the reduction of energy consumption during the production of PV systems has hardly ever been a primary development goal in PV industry. Modern large-scale continuously running module production lines, in particular for thin-film modules, should enable at least 50 % improved EPBT data.

The very high EPBTs, close to the lifetime of PV systems, which have been published in the past were caused either by inappropriate definitions of EPBT or by the assumption of obsolete module technologies (more than 400 μm thick wafers made from electronic grade silicon) or technologies for space applications.

It must be emphasized that the criterion of energy pay back, does not hold for any conventional power plant because these plants will in any case consume more energy than they will be able to supply.

What's Still an Open Question

Although LCAs of photovoltaic systems are very complex and often incomplete, no basic question in this field is open. However, current data needs to be updated with experience from new large-scale production facilities to verify the reduction potential forecast here. In addition, emissions other than the standard emissions (e.g. water emissions) and additional environmental impacts (e.g. land use) are far less well-studied. This might be of special relevance in the context of CdTe systems and the use of toxic Cd-bearing compounds. In addition, it is an open question whether the use of Cd-bearing modules will come into conflict with future and possibly more stringent regulations concerning handling and use of Cd or Cd-bearing compounds.

What Remains To Be Done

Modern and large-volume plants for the production of PV modules in particular will have to be assessed using a state of the art LCA. Generally, impact categories other than global warming and primary energy demand should also be at the focus of future LCAs. Establishing LCAs is a costly and highly complex task, but it is mandatory for PV production plants, since the acceptance of a renewable power plant technology – and a relatively expensive technology like photovoltaics in particular – depends strongly on its ecological standards, or more generally speaking, on its degree of sustainability.

For the photovoltaics industry this means

- Consumption of electricity and other energy during production has to be strongly reduced by measures such as using less semiconductor material, development of energetically optimized processes, recycling of by-products, continuous operation of production plants, thermal insulation and heat recovery.
- Economically feasible technologies for module recycling have to be developed. For CIS and Cd technology this is mandatory in any case for safety reasons. For crystalline Si technologies this could drastically reduce the specific energy demand.
- With respect to CdTe and CIS technology precautions against resource depletion have to be taken (see Sect. 4.4). An important general precaution is the diversification of thin-film technologies.

Strategies in PV Research and Development

R&D and History

A. Räuber

What's Clear

From the beginning of the development of terrestrial photovoltaic energy generation this technology has found appreciable support from governments and from industry, and this situation has not substantially changed so far. The motivation for the involvement of governments is manifold. First of all, among renewable energy options, photovoltaics have the best combination of positive features: they offer very direct one-step energy conversion from solar irradiation to electricity; electricity is energy of highest value, no mechanical or chemical processes are involved, there are no moving parts, no noise, no byproducts, and there is tremendous potential for application around the globe. Secondly, PV have the appeal of being part of the huge and rapidly growing semiconductor industry, with the promise of becoming the basis for a prosperous industry offering a large number of new jobs, not only in production but also in the field of small businesses, including installation companies.

In the 1980s and early 1990s, government activities were centered around the funding of research, development, and demonstration, whereas the late 90s and the first years of the new century are characterized by strong growth of market introduction efforts. About 1–2 billion US$ have been spent on R&D by governments up to now, and a similar amount has been invested by industry. The USA, Japan, Germany, and the European Union dominated R&D in funding for PV, but other countries also made appreciable contributions, like Italy, France, Netherlands, and Switzerland, to name just a few.

Very generally speaking, the main objective of R&D has been and still is the development of reliable technologies for the conversion of sunlight to electrical energy. In detail, this means research on materials for solar cells, development of fabrication processes for solar cells and modules, improving the product with respect to efficiency and stability, development of systems components, and most important, working on cost reduction for all parts of the chain.

Market introduction efforts have to consider two areas. First, governments must provide firm rules for the use of PV systems. In the meantime, nearly all countries now have regulations on the installation of PV systems, e.g. in buildings, on the connection of systems to an existing electricity grid, and on the obligation of the utilities to allow feeding generated electricity into the grid. The second measure is financial assistance for the investor by subsidies, cheap loans, or cost-oriented feed-in tariffs.

Japan and Germany have the largest market introduction efforts, and similar but smaller programs are found in a number of other mainly European countries like Spain, Italy, Netherlands, and others.

Regarding market introduction in the developing world, the activities of the World Bank are growing appreciably, the main activities being in India and Southeast Asia.

What's Still an Open Question

With steadily falling prices for PV modules and systems, PV becomes the first choice for electricity generation for many applications at remote sites off the power grid. However, there is need for appreciable cost reduction until PV power generation becomes competitive with current prices for electricity from the grid in industrialized countries.

For the near future, the main questions are:

- Among the many technologies studied today, which technology is best suited to fulfill the requirements of efficiency, reliability, and low cost?
- Will the potential of cost reduction be sufficient to deliver electricity at competitive prices to the consumer?
- Are governments willing to support the introduction of PV for a time sufficient to reach economic viability?

Although PV products are now on the market for more than 20 years the technology is still in its early stage.

Very promising thin-film technologies are only just entering the marketplace, but even conventional crystalline silicon technology is not really mature. It has not yet come to the mass production stage, and strong cost reductions can be expected due to scaling up production units, and also due to innovative new processing technologies.

What Remains To Be Done

As has been pointed out, there is still great potential for cost reduction of photovoltaic electricity generation. Innovative new technologies can enhance efficiency and reliability of the products, and at the same time reduce production cost.

Two actions are required to sustain effective development of photovoltaics. First, further R&D efforts are a must, since the new technologies are in their infancy in many areas. Technical developments, however, are not sufficient to reduce the cost of photovoltaics to a level that guarantees a self-sustaining market. Therefore, what is needed is the development of a true mass market with a strong up-scaling of the size of production units. This can only happen if market introduction programs continue for at least another five to ten years to offer a stable market for the young industry.

Strategies in PV Research and Development

Market Introduction Programmes

F. Staiß and A. Räuber

What's Clear

The main barriers to market introduction are still the high electricity cost of grid-connected photovoltaic systems and the lack of appropriate financing schemes for standalone systems in developing countries. To overcome them, programs other than Research & Development projects have been launched in many countries.

Market introduction programs are an important part of governmental efforts to increase the share of renewable energies in the primary energy mix. Main motivations for the introduction of "clean" energies are to protect the environment and the climate, to reduce dependence on energy imports, and to supply electricity in remote areas.

In industrialized countries, promotion programs concentrate on grid-connected systems. A number of models have been applied to facilitate financing the high initial costs. Investment subsidies and soft loans are direct measures, but special taxation measures (tax credits, enhanced depreciation) can also be very effective. Additionally, several governments have introduced enhanced feed-in tariffs for electricity from renewable energy sources like wind, photovoltaics, and others. This seems to be a very effective way to stimulate the introduction of these energy sources, since the additional cost is more or less equally distributed among all electricity customers.

Regulatory measures comprise technical codes for the safe electrical connection, building codes for the installation of photovoltaic generators on roofs and other places, and rules for the connection of these systems to the public grid.

Japan and Germany are the front-runners in the market introduction of photovoltaic systems, but a number of mainly European countries are following the route. The impact of current programs on the growth of the market and the production capacity of the photovoltaic industry is remarkably strong.

In the developing world, many countries also have taken measures to foster renewable energy technologies, and photovoltaics in particular. In most cases, programs are implemented in bilateral or multilateral cooperation with donors. In general, strategies focus clearly on small-scale rural electrification, since photovoltaics is the supply alternative that should be favored in many cases, not only for economic but also for ecological and social reasons.

What's Still an Open Question

In remote areas of the world with a weak infrastructure, most people do not have access to the public electricity grid. Thus, photovoltaics are a favorable choice for small-scale electricity supply, at least to meet basic needs like lighting, radio, telephone or water pumping. Although in many cases photovoltaics are cost-competitive with other electricity generation techniques like diesel generators, the investment lies far above the level that generally poor people can afford. Therefore, financing schemes for solar home systems (SHS) are the real need for the electrification of remote rural areas.

For developed areas with a properly functioning power grid, the situation is quite different, since without subsidies the cost of photovoltaic electricity on the spot is normally still a factor of three to five higher than the standard tariff offered by the utility. The question is whether progress in technology and introduction of mass fabrication will after some time lead to installation costs that allow the generation of photovoltaic electricity at affordable cost, and when this will happen. Probably, this will take at least another 10 to 20 years.

What Remains To Be Done

It is one of the big challenges for governments, both in developing and developed counties, to built up a sustainable electricity supply. Today, even in the most remote areas, people are aware of the benefits of modern technology, and in the future nobody will accept exclusion from using electric light and modern communication techniques. The introduction of photovoltaics is certainly the fastest way to satisfy basic needs. But only intelligent financing schemes and the engagement of big organizations will be able to provide the huge investments required.

The introduction of photovoltaic systems to generate electricity in industrialized countries in competition with the public grid is much more cost-sensitive than any other application. As has been pointed out, the introduction of real mass fabrication and the related reduction of installation costs will take a rather long time, and the long-term commitment of governments to support the introduction of this clean energy technology is needed. On the other hand, any effort in this direction will not only generate further energy technology, but also will introduce a whole new and innovative industry.

List of Contributors

Prof. Dr. Achim Bubenzer
Fachhochschule Ulm
Ulm University of Applied Sciences
Prittwitzstr. 10
89028 Ulm, Germany
bubenzer@fh-ulm.de

Prof. Dr. Joachim Luther
Fraunhofer Institut
für Solare Energiesysteme
Oltmannstr. 5
79110 Freiburg, Germany
luther@ise.fhg.de

Dr. Hermann Maurus
Angewandte Solarenergie-ASE GmbH
Division Phototronics
Hermann Oberth Str. 11
85640 Putzbrunn, Germany
hermann.maurus@solar.rwesolutions.com

Dr. Joachim Nitsch
Deutsches Zentrum für Luft- und Raumfahrt e. V. (DLR)
70569 Stuttgart, Germany
Joachim.Nitsch@dlr.de

Martin Pehnt
Ifen-Institut für Energie- und Umweltforschung GmbH
69120 Heidelberg, Germany
Martin.Pehnt@ifen.de

Dr. Klaus Preiser
Fraunhofer Institut
für Solare Energiesysteme
Oltmannstr. 5
79100 Freiburg, Germany
Preiser@ise.fhg.de

Dr. Armin Räuber
PSEmbh
79114 Freiburg, Germany
raeuber@pse.de

Dr. Horst Schade
Angewandte Solarenergie-ASE GmbH
Division Phototronics
Hermann Oberth Str. 11
85640 Putzbrunn, Germany
horst.schade@solar.rwesolutions.com

Dr. Thomas Schott
Zentrum für Sonnenenergie
und Wasserstoff-Forschung (ZSW)
Baden-Württemberg
70563 Stuttgart, Germany
thomas.schott@zsw-bw.de

Dr. Frithjof Staiß
Zentrum für Sonnenenergie
und Wasserstoff-Forschung (ZSW)
Baden-Württemberg
70563 Stuttgart, Germany
staiss@zsw-bw.de

Dr. Gerhard Willeke
Fraunhofer Institut
für Solare Energiesysteme
Oltmannstr. 5
79100 Freiburg, Germany
Willeke@ise.fhg.de

1 Strategies for Introducing Renewable Energies and the Contribution of Photovoltaics

J. Nitsch and F. Staiß

1.1 Development of the World Photovoltaic Market in the Past

1.1.1 Development of Module Production and Prices in the Past

The market for photovoltaic (PV) systems has undergone very dynamic development over the last two decades. Annual production has grown from 5 MW_p in 1980 to 280 MW_p in 2000 (Fig. 1.1). In the 1990s production quadrupled, which corresponds to an average growth rate of 17 %/yr. In the last three years, more than 30 %/yr has been reached [PV Report]. In total, more than 1 100 MW_p PV capacity has been installed worldwide by the end of 1999.

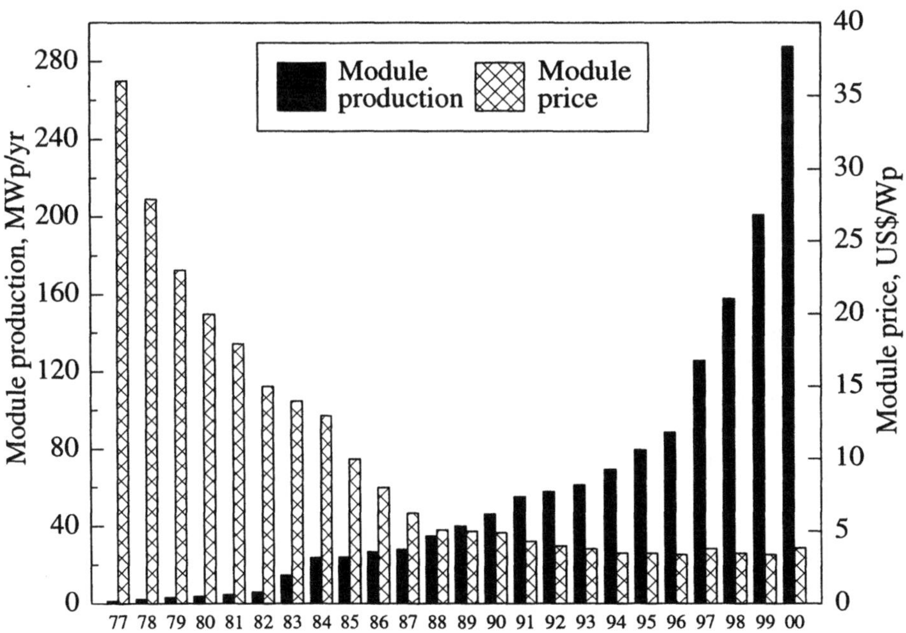

Fig. 1.1. Development of annual global PV module shipments and prices since 1977

The driving forces behind this development are related to the specific characteristics of PV compared to other energy technologies:

- *No resource limitation:* solar energy potentials exceed by far the world's energy demand.[1] In comparison to other resources, solar energy is most homogeneously spread over the globe.
- *Extreme modularity and large spectrum of applications:* due to the modularity of PV, it is possible to realize systems from a few milliwatts to the multi-megawatt range.
- *Low maintenance:* PV systems usually work without any moving or thermally stressed parts. Thus, they require only a minimum of maintenance.
- *Environmental benignness:* during operation, systems neither emit any noise nor any pollutants. Energy requirements for the production, operation, and recycling of present PV modules are considerably less than the electric output over the lifetime, and will further decrease in the future. The risk of hazards to the environment or to human health in case of accidents is low, due to the absence of a fuel cycle.
- *Independence of site:* since solar energy is available at any site and PV systems can be operated as standalone systems without any infrastructure, they offer the highest flexibility in site selection.
- *Well-known to the public:* PV is well-known to the public since there are a large number of PV products that can be used by anybody (e.g. pocket calculators, watches, lamps, camping equipment).
- *Economical electricity costs in small and smallest applications:* although PV counts as rather expensive, it is nowadays economically competitive in a number of market niches, especially in standalone applications in the power range below $1\,\mathrm{kW_p}$.
- *Large-cost reduction potentials:* PV is still an expensive technology, but is also in an early phase of technological and industrial development. Thus, large cost-reduction potentials can be exploited, increasing competitiveness over a wider scope of applications.

By the early 1980s, these outstanding characteristics motivated public institutions to support the transfer of PV from research laboratories to practical applications. The financing of a number of demonstration projects in combination with intensified research and development efforts paved the way for entering first market niches. Today, commercial applications, which are independent from public subsidies, can be estimated to be between 50 % and 60 % of the total market.

The growing demand for PV systems on the one hand and technical developments on the other have stimulated each other, and led to considerable cost reductions while technical performance has improved. As Fig. 1.1 shows, prices (not production costs) for PV modules have fallen by a factor of 3 (in real prices) since the early 1980s. Today, factory prices are in the range 3.5 to

[1] For details see Sect. 1.2.2

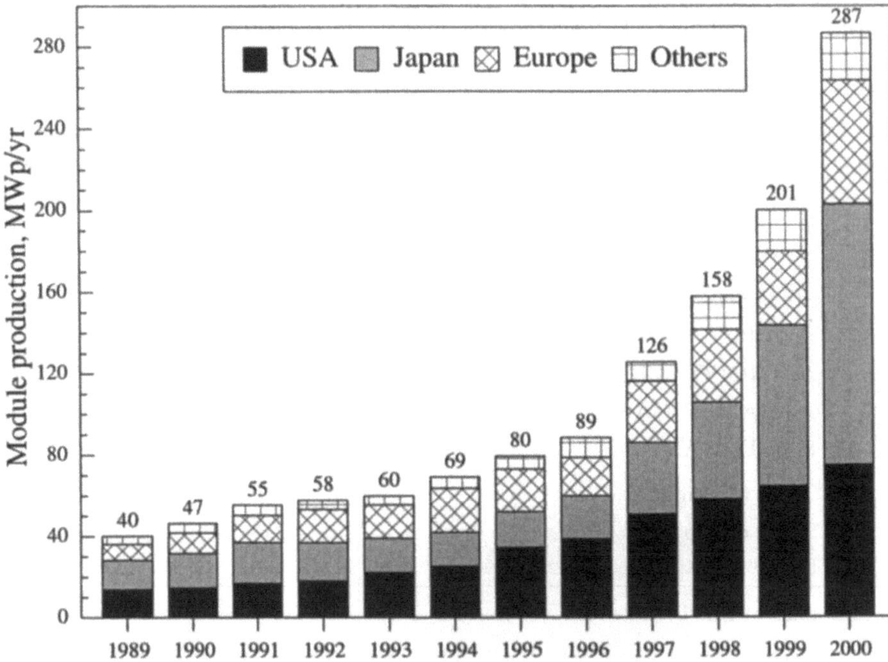

Fig. 1.2. Market shares of the most important regions manufacturing PV modules during the last decade

4 $/$W_p$ (prices for retail consumers are usually higher). The leading countries in PV manufacturing are Japan (market share 45 %) and the United States (25 %), while all European countries together make up 20 % [PV Report] (Fig. 1.2). The remaining 10 % fall to India, Australia, Taiwan, and China, although the third-world market is rather important. Consequently, international trading of PV modules is common practice. American companies, for example, export about three quarters of their production.

It is worthwhile to note that in contrast to other renewable energy technologies, PV module production is in the hands of a small number of big industrial companies which can afford high investments in production facilities, and can rely on a global distribution network. They produce PV modules under their own names or stand behind companies as major shareholders. As Table 1.1 shows, the top ten producers account for 80 % of the world market.

1.1.2 Development by Applications

The commercial non-subsidized PV market almost exclusively comprises decentralized applications with low electricity demand [PV 2010 1995] (Fig. 1.3), such as the so-called *consumer market* (e.g. pocket calculators, watches, radios) with capacities between some milliwatts and approximately

Table 1.1. Ranking of PV cell/module manufacturers in 1999 [PV News 2000]

Rank	Company	Countries of production	Production 1999 (MW_p)
1	BP/AMOCO SOLAREX	USA (18.0), Australia (5.5), Spain (5.0), India (4.0)	32.5
2	KYOCERA	Japan	30.3
3	SHARP	Japan	30.0
4	SIEMENS SOLAR[a]	USA (26.0), Germany (2.0)	28.0
5	SANYO	Japan	13.0
6	ASTROPOWER	USA	12.0
7	ASE AMERICAS	USA (4.0), Germany (7.0)	11.0
8	PHOTOWATT	France	10.0
	Total top eight		166.8
	Share of top eight of total production		83 %

[a] now part of Shell

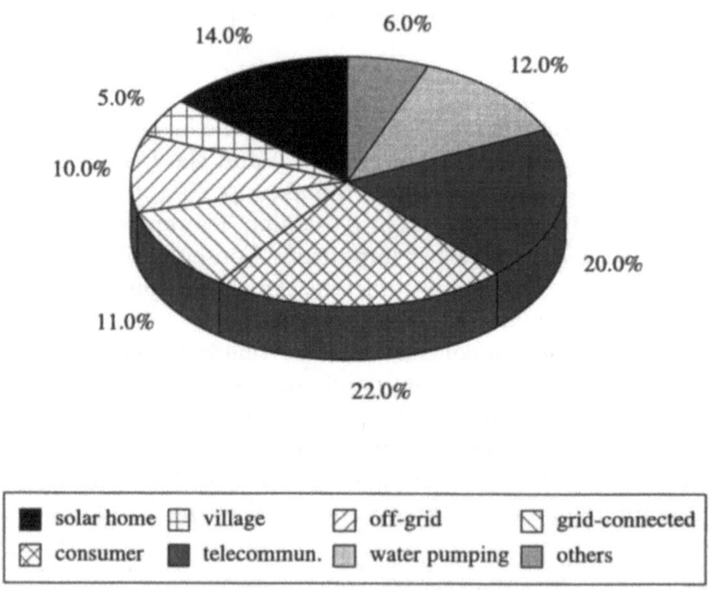

Fig. 1.3. Structure of global PV market by applications (period 1990–1994)

1 W_p, and the *leisure market* (camping, hobby, garden) with capacities up to 50 W_p. In both market segments more than 20 % of the total module production today is sold. The consumer market is the domain of amorphous silicon solar modules, since they are cheaper than crystalline silicon modules and the question of high module efficiency and long lifetime is of minor importance. The consumer market roughly absorbs half of the annual amorphous silicon solar module production (total 1998: 20 MW_p). The other half is sold to the leisure market.

Another important market segment can be called *small PV systems for professional applications*. It comprises telecommunications (e.g. relay stations), cathodic protection of pipelines, military applications, signaling equipment, parking ticket machines, and others. This market segment also absorbs about 25 % of total module production, but in contrast to the consumer market, requires technically reliable systems with a long lifetime.

The fourth big market segment comprises typical applications in *developing countries*. Here, most relevant are so-called *solar home systems*, which are standardized PV systems from 10 W_p (so-called solar lantern systems) to a maximum 100 W_p to produce electricity for lighting and the operation of radio/TV sets. In many regions of the world, solar home systems are the cheapest (and often the only) way to provide electricity for consumers in rural areas. Consequently, there is great demand in a number of countries: today more than half a million PV systems are installed, making up 14 % of the PV world market [Posorski 1998].

Apart from solar home systems, *water pumping systems* (drinking water and irrigation) and *village electrification systems* (e.g. communal buildings, medical posts, schools) are typical applications in countries with weak infrastructure in rural areas. Including the market segment of off-grid systems, they account for another quarter of total module production. Electric power ranges between 100 W_p and some kilowatts. The systems are on the threshold of economic competitiveness, not only against grid extension but also against small diesel generators.

Another big market segment are *grid-connected PV systems* which comprise systems on buildings in the power range 1 to 10 kW_p, and solar façades and larger installations in open spaces with capacities of some 10 kilowatts up to a few megawatts. The biggest plants ever erected up to now are Carissa Plains/USA (6.5 MW_p) and Serre/Italy (3.3 MW_p). For a few years now, even larger systems up to 50 MW_p have been discussed, but have not been realized yet (e.g. Greece or India). Although grid-connected PV systems today are not competitive without subsidies, especially in Europe, the United States, and Japan, a considerable amount of money was invested in such systems (an estimated $250 million in 1997). While large demonstration systems in the technologically leading countries serve as a showplace to a certain extent, the strategy in the field of rooftop systems is to reach competitiveness against conventional electricity supply of retail consumers step by step. This is especially the case in the sunny regions of the United States and in Japan, where high electricity prices for private households are the driving force.

To sum up, it can be estimated that the international PV industry now generates in 2000 an annual turnover in the range of \$3.0 billion per year, of which about half relates to PV modules.

1.1.3 Development by Regions

In the 1990s, about 60 % of the world PV module production was installed in industrialized countries, and 40 % in the developing world [IEA 1998] (Fig. 1.4). Larger grid-connected systems have been installed almost exclusively in the USA (50 % of total installed capacity in this segment), Germany (18 %) and Italy (17 %). The situation is similar for building integrated systems, but in this case the ranking is Japan (44 %), Germany (30 %) and USA (15 %). In the field of PV standalone systems, developing and newly industrialized countries account for the biggest portion.

In *industrialized countries* the installed capacity more than tripled between 1992 and 1998 from 92 MW_p to 320 MW_p, which corresponds to an average increase of 25 %/yr. The leading countries in absolute terms are Japan, with a total installed capacity of 133 MW_p by the end of 1998, USA with 108 MW_p, and Germany with 54 MW_p, which accounts for three quarters of

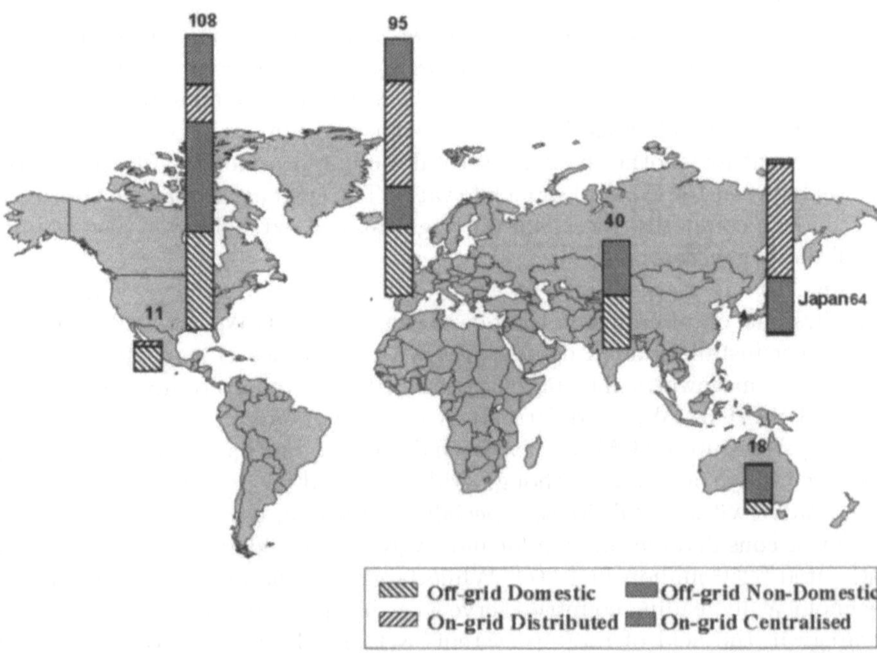

Fig. 1.4. Installed PV capacity in nine selected countries by the end of 1998

total capacity. It is notable that the ranking changes if these figures are related to the population in the respective country. With values of more than 1 W_p per capita, Switzerland, Austria, and Japan are the countries leading by far. For Germany it is around 0.7 W_p, while the values for all other nations are lower. Although more than half of the PV capacity is still used for off-grid applications, small distributed on-grid applications (rooftops) especially have strongly increased by close to 1 000 % from 13.5 MW_p to 147 MW_p. Obviously, this is strongly related to national policies. On-grid systems, including large grid-connected systems, account for more than 60 % of the total capacity only in Germany, Japan, Switzerland, and Denmark (Fig. 1.5).

The value of PV business in the countries indicated in Fig. 1.5 exceeded $1 billion in 1997, which is 30 % more than in 1996. The biggest markets are the United States whitch accounts for $410 million, Japan with $230 million, Spain with $160 million and Germany with $110 million. The business accounts for to 7 600 employees, with 3 700 working in the manufacturing of PV-specific components, and 1 600 in R&D (government agencies, consultancies, but not R&D at manufacturers). The remaining 2 300 jobs have been created in utilities and distribution/installation companies.

The market in *developing countries* has grown in a number of countries. The most widespread systems are solar home systems, the figures for which range from more than 500 000 to more than 700 000 [Posorski 1998], telecommunication systems (about 200 000), water pumping systems (about 50 000), and systems for medical and educational applications (about 10 000).

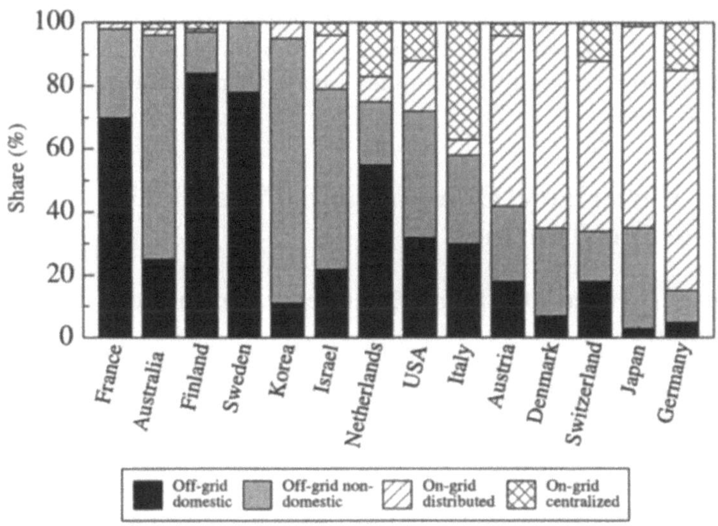

Fig. 1.5. PV capacity by application in 14 selected IEA countries by the end of 1998 (total 365 MW_p)

Table 1.2. Relevant PV markets in developing countries

Country	Installed PV systems[a]	Considered Systems
India	450 000	SHS, street lights, water pumps
Kenya	150 000	SHS
China	100 000	SHS
Zimbabwe	85 000	SHS
Morocco	70 000	SHS
Indonesia	60 000	SHS
Mexico	40 000	SHS
Sri Lanka	10 000	SHS
		SHS = solar home systems

[a] Estimates for 1998

Most of the installations have been implemented not only with support from international organizations like the World Bank, the Global Environmental Facility, regional development banks, and national aid institutions, but also from the PV industry (e.g. the US Export Council for Renewable Energy). Relevant markets are indicated in Table 1.2, the biggest of which is India where systems amount to a total capacity of 40 MW_p altogether.

In 1997 alone, more than 200 000 systems were added, which corresponds to 10 % of annual PV module production. It can be estimated that the annual turnover in this market segment is on the order of $100 million. In developing countries, probably more than 100 000 families receive their income from solar business [Cameron 1998].

1.2 Present Electricity Demand and Potential of Renewable Energies

1.2.1 Present Structure of Global Electricity Production and Demand

The present contribution of renewable energies (RES) to the global energy supply makes up 18 % (Table 1.3) concentrating on the use of hydro power and biomass. The relatively large share of biomass, however, is mainly caused by the non-sustainable use of firewood in the poorer developing countries (non-commercial energy), and is thus no basis for a sustainable application of biomass. Only one tenth of the present biomass share can be regarded as ecologically tolerable, including energetical exploitation of residual wood and organic waste (mostly in the industrialized countries), as well as generation of biogas from keeping livestock and from food production. Considering this

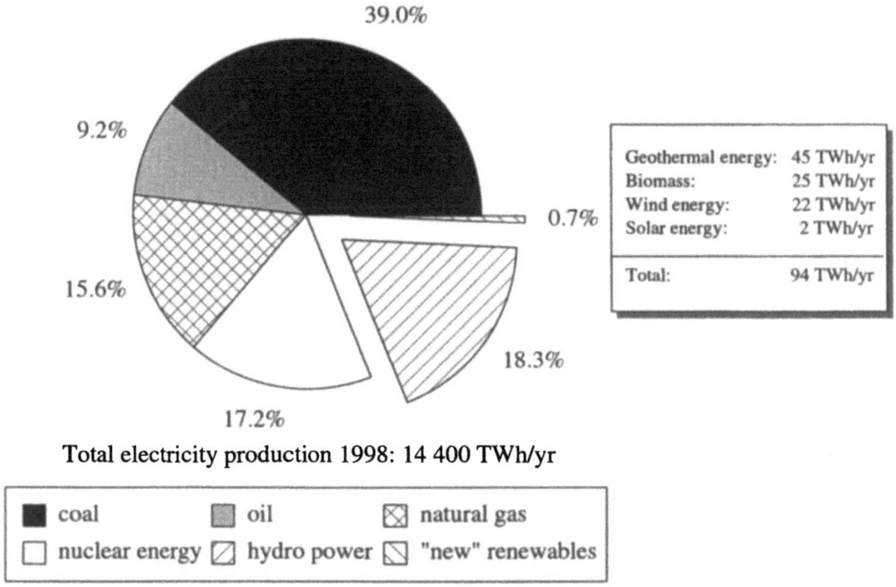

Fig. 1.6. Present global electricity production and share of primary energy sources

fact, the value to be taken as a basis for sustainable growth of RES is about a 5 % share in the present global final energy consumption.[2] The real "modern" technologies for using energy from solar irradiation and wind as well as geothermal heat, however, only contribute a mere 0.4 % to the coverage of the present worldwide final energy demand.

This fact becomes especially obvious in the structure of global electricity production (Fig. 1.6, [IEA 1999], [UN 1998], [Wind 1999]). Coal (hard coal and lignite) is the most important energy source, while hydro power rates second with about 700 GW and a share of 18 % in the overall electricity production, which amounts to 14 400 TWh/yr (1998).

All the remaining RES, however, only supply 0.7 % of the electricity produced worldwide, with 8 000 MW from geothermal power stations, 5 000 MW from biomass power stations and cogeneration plants, 10 000 MW from wind power plants, 350 MW from solar thermal power plants, and about 900 MW from photovoltaic plants, which is altogether 94 TWh/yr, with solar irradiation contributing the smallest share, 2 TWh/yr. Global electricity production by RES with its total of 2 700 TWh/yr (98 % of which is from hydro power) is all in all higher than that of nuclear energy. At present, wind energy shows extraordinarily high growth rates, with about 3 000 MW/yr (i.e. 1 600 MW in Germany alone in 1999).

[2] This is 9 % of the global primary energy consumption in 1997, with 395 EJ/yr [IEA 1999].

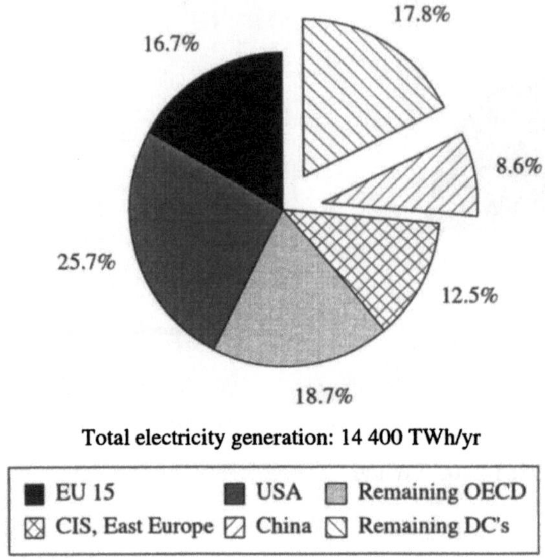

Fig. 1.7. Structure of present (1998) global electricity consumption by regions

Presently, there are particularly big differences between industrialized and developing countries in their electricity consumption, by a factor of 10 (Fig. 1.7), with the industrialized countries consuming almost 75 % of the electricity produced. The average per capita figures are for the OECD 8 600 kWh/yr (USA 14 700; EU 7 250), for the CIS and Eastern Europe 3 750 kWh/yr, whereas the average figures for China are only 980 kWh/yr, and 750 kWh/yr for the remaining developing countries. Even with great potential for a more rational use of electricity, which is indeed possible with up-to-date appliances and drives, and further technical progress, a considerable demand of the developing countries to catch up in the electricity sector can be expected.

1.2.2 Potentials of Renewable Energies

Considerations on using renewable energy sources are based on their extraordinarily great physical potential and the fact that unlimited energy sources which already exist in the ecosphere can be used for human purposes. Thus the most crucial criteria for ecological sustainability can be fulfilled. The solar irradiation on the continents, the potential energy of water, the kinetic energy of wind, waves and tides, the chemical energy stored in the permanently growing biomass, the thermal energy of the oceans and the geothermal energy annually offer about 3 000 times as much energy in the form of unlimited energy flows (Table 1.3, column 2) as is presently consumed worldwide. Based on these data, the *technical potentials* describe the energy, which can be supplied from the present point of view, in a form the final consumer can make use of, i.e. direct heat of different temperatures, electricity and fuels for heat generation and transportation (e.g. hydrogen) produced with RES. When determining these technical potentials, a number of restrictions have to be considered:

- potentials and limits of technological efficiencies, possible capacity of a plant, technical potentials for development of the technologies for energy conversion and use which are available already, or which will be available in the near future
- structural restrictions for exploitation due to the dependency on sites (e.g. geothermal energy), limited transportation radii (e.g. biomass), availability of area or its competitive use (e.g. collectors, solar panels, cultivation of energy crops), limited availability and reliability of energy supply (e.g. electricity from fluctuating sources), discrepancies in demand and supply (e.g. surplus that cannot be used or excessive peak loads with wind and solar electricity)
- ecological restrictions with respect to the area required (e.g. wind energy), disturbance of flowing waters (e.g. hydro power), limited possibilities for the use of biomass (e.g. residual material from forestry and agriculture; cultivation of energy crops)

Statements on the technical potential for the use of renewable energies therefore depend on numerous assumptions and are not to be regarded as static. If these criteria are taken into consideration, merely a share of *some thousandths* (irradiation, wind, energy of oceans) to *some hundredths* (biomass, geothermal heat) of the physical energy flows mentioned can be used energetically, i.e. in the form of usable secondary energy carriers. Hydro power already "concentrated" is an exception where technical exploitation of around 10 % is possible.

Making technical use of the physical energy flows to such an extent, however, would be enough to cover the entire energy demand of the whole population, in principle, even if requirements still continue to rise (Table 1.3, column 3), because from the technical point of view at least *six times* as much energy (final energy) as is presently used could be supplied by energy carriers without any restrictions. For this estimate the area for making use of solar irradiation has quite restrictively been limited to 2 % of the global land surface, which is about 10 % of the area *not* covered by settlements, forests, and agriculture at present. Obviously, the area could be even larger. The potential areas for growing energy crops make up about 5 % of the area used for agriculture.

The following data are to be taken as a further example for the global technical potential of RES only in the field of electricity generation with wind, water, and solar thermal power plants [Wind 1999], [IEA 1999], [World Atlas 1998], [Klaiss 1992]:

- wind power, land-based = 53 000 TWh/yr [Wind 1999]
- hydro power = 14 300 TWh/yr electricity [World Atlas 1998]
- potential of solar thermal power plants in suitable areas in the Mediterranean region = 350 000 TWh/yr [Klaiss 1992]

These figures, which together add up to more than 30 times the global electricity production of 1998, show that supplying the world's population almost entirely with RES is not at all endangered because of limits fixed by their potential.

To supplement line 1 of Table 1.3, the technical potentials of photovoltaics are explained in more detail in Table 1.4. Basic values are, on the one hand, the global land surface (149 million km^2), and on the other hand the land area of Germany (357 000 km^2). Certain selected areas with high solar irradiation, which are not used at present and are suitable for building large solar plants [Winter 1988], complete this table. The annual amounts of global irradiation range from 900 kWh/m^2yr in southern Scandinavia to a maximum of 2 500 kWh/m^2yr in the Central Sahara Desert and the US dessert southwest.

Table 1.3. Annual global potential and approximate technical potential of renewable energy sources (based on global final energy consumption in 1997 = 1), [TAB 2000]

Energy source	Total physical supply	Technically usable (final energy)[a]	Presently used (final energy)
Solar irradiation	2 850[b]	3.80[c]	0.001
Wind power	200	0.05[d]	0.0003
Biomass	20	0.40[e]	0.140[f]
Geothermal heat	20	1.00	0.003
Energy of oceans[g]	10	0.05	0
Hydro power	1	0.15[h]	0.035
Total	3 100[i]	5.90	0.180

[a] Final energy consumption (1997 = 9.4 billion t coe/yr = 275 EJ/yr [IEA 1999])
[b] Irradiation on continents (= 15 % of total solar irradiation)
[c] 3 million km^2 surface for collection (= 2 % of the global land surface) with 40 % direct heat, 40 % electricity and 20 % hydrogen (for example)
[d] Only land-based; mean value from [WEC 1995], [Grubb 1993]
[e] Residual material from forestry and agriculture and 1.5 km^2 area for cultivation (= 1 % of global land surface)
[f] Present consumption of firewood and organic waste
[g] Energy from waves, tides, temperature gradients
[h] "Technically usable" according to [World Atlas 1998]
[i] Global annual energy consumption corresponds to 3 hours of physical potential of RES

Table 1.4. Potentials of electricity production with photovoltaics

	Worldwide (all continents included)	Selected areas with high irradiation[a]	Germany
A) Solar irradiation			
Land area, million km^2	149	1.88	0.357
Total irradiation on horizontal surface, 1 000 TWh/yr	235 000	4 050	357
Specific annual irradiation, kWh/m^2yr	1 575	2 150	1 000
B) Maximum technical potential, TWh/yr			
Usable area, % of total land area	1.0	50.0[a]	1.0
Electricity (mean annual efficiency: 10 %)[b]	235 000	200 000	357
Share of present electricity consumption	16	14 (worldwide)	0.7
Electricity (15 %)[c]	350 000	300 000	535
Share	24	21	1.0
C) Practical technical potential, TWh/yr			
Usable area, % of total land area	0.1	5.0	0.28[d]
Electricity production (15 %)	35 000	30 000	150
Share of present consumption	2.4	2.1	0.3
Installed Power, GW_p	26 500	17 000	160
Scenario values (in 2050), TWh/yr			
Upper value (see Chap. 1.3)	2 500	900	40
Share of practical technical potential, %	7	3	27
Corresponding installed power, GW_p	1 900	500	43

[a] Selected areas especially suited for large solar plants (e.g. global radiation $> 2\,000\ kWh/m^2yr$; homogeneous small-shard rock deserts, no danger of sand storms or heavy rainfall; currently not used, little or no vegetation. Distribution: North Africa (Sahara Desert) 45 %, Arabian Peninsula 23 %, Australia 12 %, Central Asia 9 %, America 8 %, South Africa 3 %. Possible factor for land use: 0.50 (module surface/land surface). Source: [Winter 1988]

[b] Annual mean system efficiency with present technology

[c] Annual mean system efficiency with future technology

[d] Roofs, façades, other surfaces within areas of settlement. Source: [BMU 1999]

The corresponding duration of sunshine hours ranges between 1 500 h/yr and 3 400 h/yr.[3] For Germany 1 000 kWh/m^{2}yr and 1 600 h/yr can be taken as representative values. If a maximum 1 % of the land surface is considered to be usable area for PV modules, with present technology (median annual system efficiency = 10 %) electricity consumption 16 times the present could be covered worldwide. As far as the figures go, this value could be produced solely in areas with high solar irradiation determined in [Winter 1988].

In Germany, as an example of a densely populated industrialized country with high specific energy consumption situated at medium latitude, about 70 % of the electricity demand could be supplied if 1 % of the land surface were used. In the long term, however, systems efficiencies of around 15 % can be assumed, which increases contributions correspondingly and comes approximately up to the present total electricity consumption in Germany.

Considerations on incorporating PV systems practicably and in an economically reasonable way into existing settlement structures, i.e. by using for example roofs, façades, parking lots, noise barriers, and other installations within settlements, as well as consideration of the competitive potential of thermal collectors for hot water generation and room heating, suggest assuming an avialable surface share of approximately a mere 0.1 % of the total land area. Furthermore, adding the resources of large plants on 10 % of the areas with high irradiation would lead to 65 000 TWh/yr (43 500 GW_p, 243 000 km^2 of module surface), which is more than *four times* present global electricity consumption, as a value for orientation for the practical technical potential of PV electricity generation. In Germany approximately 1 000 km^2 of module surface (= 0.28 % of land area) could be used for photovoltaics if these criteria were adhered to. As a result there would be an electricity generation potential of 150 TWh/yr or 30 % of present consumption [BMU 1999].

These very limited technical potentials are still considerable enough to be exploited only to a relatively low degree in the short term, i.e. within the next few decades. Even optimistic scenario values, which are deduced in the following chapters and which presume a contribution of photovoltaics to (increasing) global electricity consumption of about 10 %, will have used only approximately 7 % of the global decentralized potentials and 3 % of the potential for centralized large plants by 2050, if a distribution of 75 % and 25 % to these categories is assumed as an example. In several individual countries (e.g. Germany) the degree of exploitation, however, will have exceeded 20 % by that date due to smaller potential values.

[3] Number of hours with more than 120 W/m^2 of global irradiation

1.3 Global Prospects for Renewable Energies

1.3.1 Scenarios for Future Global Energy Demand and Generation

Numerous studies on worldwide future energy consumption assume that RES will contribute considerably by the middle of the next century (Fig. 1.8; [Shell 1995], [WEC 1995], [WEC 1998], [Johansson 1993], [Lovins 1999], [Nitsch 1999]), regardless of the other assumptions made in these studies on the effect of a more rational use of energy and the contribution of nuclear energy. The probable contributions of RES will reach up to 20 billion t coe/yr (Shell scenario), which is more than the worldwide total energy consumption at present, and will make up 25–75 %[4] of the total supply. In view of their po-

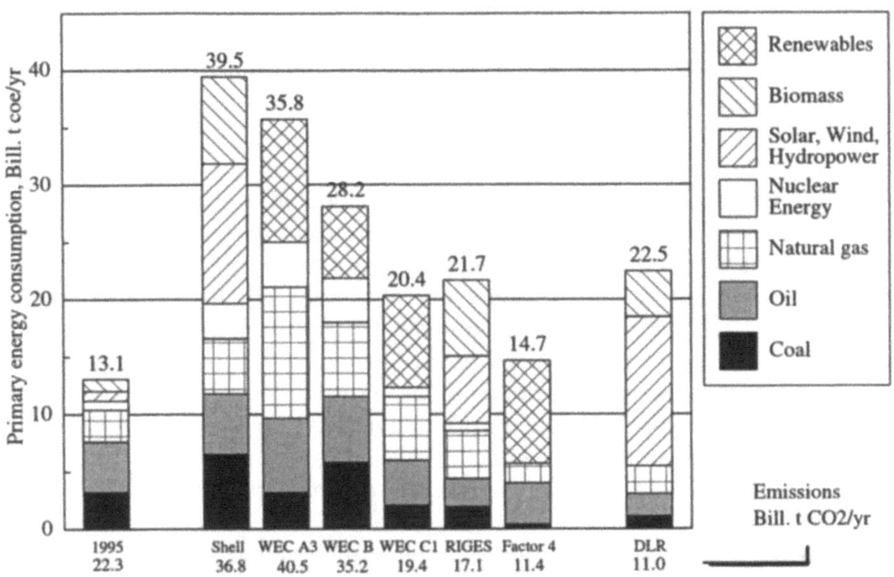

Fig. 1.8. Different scenarios for global energy consumption by 2050 and comparison with present primary energy consumption; Shell = "Sustainable development" scenario [Shell 1995]; WEC = World Energy Conferences 1995 and 1998; RIGES = "Renewable-intensive global energy scenario" [Johansson 1993]; Factor 4 = Scenario from [Lovins 1999]; DLR = our calculations [Nitsch 1999]; 1 billion t coe/yr = 29.3 EJ/yr

[4] Despite high RES contributions and increasing contributions of nuclear energy, scenarios assuming traditional growth rates in the industrialized countries and transferring it to developing countries as well *do not achieve any reduction* in global CO_2 emissions (Shell, WEC A, WEC B). This stresses the enormous importance of much more rational worldwide energy use as is assumed in the WEC C, RIGES, and DLR scenarios, and especially in the "Factor 4" scenario.

tential, RES can quite well be imagined to make such contributions according to the above considerations. From the present point of view, however, it is not clear how the necessary global growth dynamics of RES can be launched as punctually and extensively as required to have the assumed contributions realized by the middle of the next century already.

Probably the most extensive analyses in this respect have been carried out by the World Energy Council (WEC), and were last discussed in Houston in 1998 [WEC 1998]. In general, the WEC sets up several energy scenarios in certain groups (as a rule each group contains several scenarios): group A describes a process which is strongly oriented toward strong economic growth and plenty of available energy. Group B is often taken as a trend scenario ("business as usual"), whereas group C describes so-called "ecologically driven" scenarios. In more recent studies on scenarios [WEC 1998], the WEC has clearly corrected former misjudgements with respect to enormously increasing worldwide consumption of primary energy. The values assumed for energy consumption are now considerably lower and thus cast the feasibility of replacing fossil energies with RES in quite a different light. Thus scenarios with intensive rational energy use, as shown in the right-hand section of the figure (RIGES, Factor 4; DLR), also become a point to be considered together with the WEC studies.

The future importance of RES within the global energy system becomes clear especially from the description of future electricity generation and the mix of energy sources used for electricity generation (Fig. 1.9). Apart from "Factor 4" with very high rational use of energy, all the scenarios assume strong increases in electricity consumption, particularly in developing countries. The differences in the increases are generally due to differing assumptions on rational use of electricity in the industrialized countries, which already have high per-capita consumption. There is also a very large increase in the industrialized countries assumed in the prognosis of the IEA [IEA 1998], where electricity consumption by 2020 is assumed to have reached almost twice the amount of 1998, with the share of RES decreasing at the same time. In the other scenarios, however, the absolute contribution of electricity generation with RES by 2020 is assumed to be twice (scenario B1, C1) or three times (scenario A1, DLR) the amount of 1998, which leads to shares in the total production of 25–35 %. According to the WEC scenarios of 2050, electricity supply with RES will then be four times (scenario B) to six times (scenarios A1, C1) the present amount. In scenarios with high electricity consumption, such an increase brings about RES shares of about 35 %, and 58 % in scenario C1.

With the C1 scenario, the WEC has for the first time presented a path for development which proves that a risk-minimizing and sustainable energy strategy is possible worldwide. According to this scenario it is possible both to reach decisive goals for climate protection in the 21st century (limiting the increase in CO_2 concentration to less than 450 ppm and the global increase in

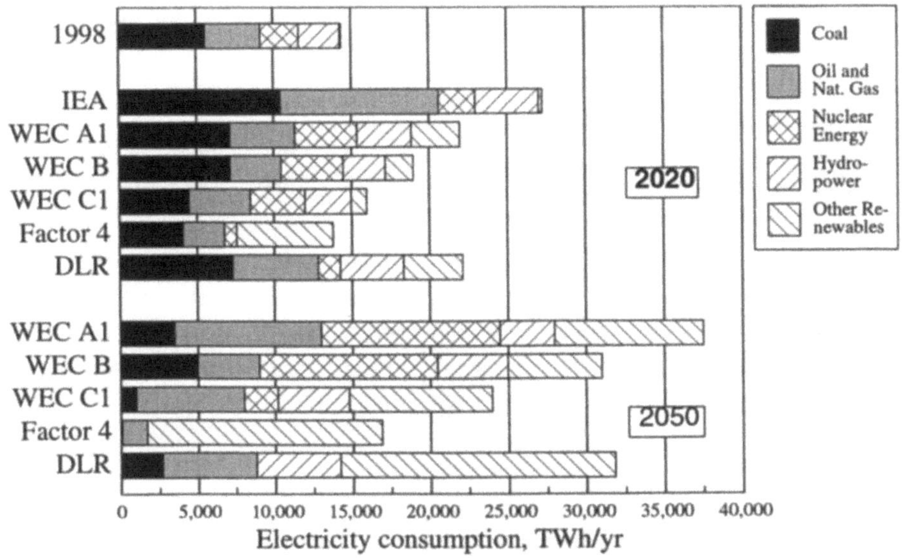

Fig. 1.9. Scenarios of global electricity consumption 2020 and 2050 from [IEA 1998], [WEC 1998], [Lovins 1999] (Factor 4) and [Nitsch 1999] (DLR) and structure of primary energy sources

temperature to less than 2 °C compared to the pre-industrial status) and to do without nuclear energy worldwide in the long term (i.e. according to WEC by 2100). However, a sufficient reduction in CO_2 (by 2/3 compared with 1990) will only be achieved quite some time after 2050, while the German climate Enquête commission and the Intergovernmental Panel on Climate Change (IPCC) call for a worldwide reduction in CO_2 by 50 % [Enquête 1995] to be effected by 2050.

All the given paths for development are considered feasible by the WEC. Especially with the C-group scenarios it is a fact that the developments will not set off as a matter of course, but that there is considerable pressure for action in energy politics. The WEC, however, has shown that risk minimization and climate protection are possible and can surely be financed, but the authors stress that in spite of the long period considered, decisions on the kind of future energy development to be pursued have to be made right now, as the strategies depicted in the scenarios will rule each other out within a couple of decades. Also, the periods for capital binding in the energy system are usually several decades. A scenario which fulfills the claims of the IPCC for 2050 with regard to CO_2 emissions and at the same time considers the needs of developing countries to catch up in energy was described in [Nitsch 1999] (see Fig. 1.8, right-hand bar). The total contribution of RES to global consumption of primary energy will be 75 % in 2050, but with its 17 billion t coe/yr is lower in absolute terms than the contribution of almost 20 billion t coe/yr in the Shell scenario (Fig. 1.8, left-hand bar). The scenario

Table 1.5. Energy shares of global electricity generation in the long-term "Solar Energy Economy" scenario from 2010 to 2050 (TWh/yr)

Energy source	1998	2010	2020	2030	2040	2050
Coal	5 621	7 310	7 350	6 450	4 800	2 700
Oil, natural gas	3 570	4 883	5 500	5 800	6 000	6 100
Nuclear	2 480	2 100	1 460	770	0	0
Hydro	2 635	3 555	4 100	4 500	4 900	5 400
Other renewables	94	752	3 740	7 980	12 900	17 600
Total generation	14 400	18 600	22 150	25 500	28 600	31 800
Total renewables	2 729	4 307	7 840	12 480	17 800	23 000
Share of RES, %	19	23	35	49	62	72
Share of RES without hydro, %	0.7	4	17	31	45	55
RES growth rate (without hydro), %/yr		17	16	7.5	4.8	3.1

describes the goals of sustainable energy supply for the next century, which are protecting the climate, minimizing risk, and bringing into line the global living conditions in an ideal way. It shows a possible transition from the present status towards this goal, and can therefore be regarded as an orientation for the necessary steps to be taken in politics, economics, and science in order to solve the problems that have arisen.

The corresponding scenario for global electricity supply (Table 1.5, Fig. 1.10) assumes very efficient use of electricity in the OECD countries, with a 30 % reduction in per capita electricity consumption by 2050 and a considerable increase in fast-developing and newly developing countries, with per capita electricity consumption to have tripled by 2050. With its total electricity production of 31 800 TWh/yr assumed in 2050, i.e. 2.2 times the value in 1998, the scenario is approximately in line with the WEC's B scenario in Fig. 1.9 with respect to consumption, but as far as the structure of electricity generation is concerned it is quite similar to the WEC's C1 scenario. At present, coal is the most important energy carrier for electricity generation. with a share of 40 %, which in itself causes about 25 % of global CO_2 emissions (and 85 % of the total emissions caused by electricity generation). Coal will still play this part over the next 30 years in spite of great efforts to mobilize the use of RES. Producing electricity with gas will become more important and will, over the long term, replace coal as the leading fossil energy carrier. The goal of risk minimization leads in the scenario to a steady decrease nuclear energy, and to its application coming to an end by 2040.

The demands made on RES contributions to the future global electricity supply and the resulting growth rates can be derived from the above goals of

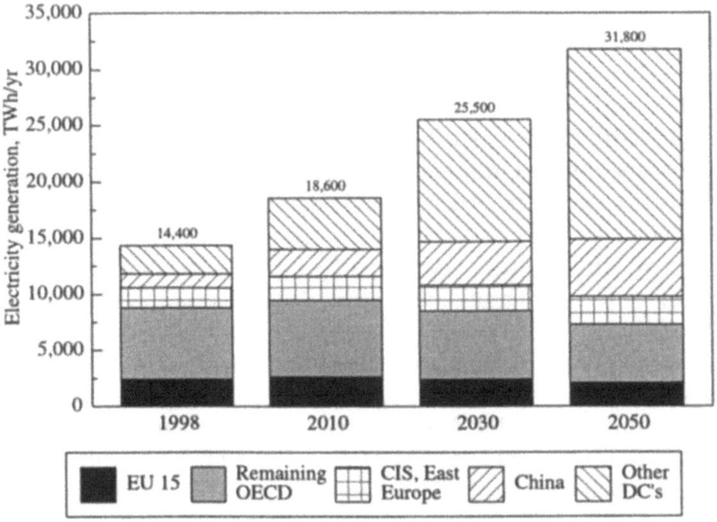

Fig. 1.10. Global electricity generation by regions within the long-term scenario "Solar Energy Economy" from 2010 to 2050 [Nitsch 1999]

the scenario. Due to high increases in consumption in developing countries and the assumed decrease of nuclear energy use, these demands are very strict. Thus, in the following, these demands are assumed in order to deduce an upper bound on the potential contribution of photovoltaics and on the market growth required in this respect. For a timely and adequate mobilization of RES (without hydro power) in the sense of this scenario, their share has to increase by 2020, with high growth rates averaging 16–17 %/yr. RES will then cover one third of the electricity demand. Then growth rates of less than 10 %/yr, dropping to 3 %/yr between 2040 and 2050, will be enough to make RES the dominant energy source of the next century, with a share of about 70 % in 2050.

1.3.2 Global Contribution of Renewables in Detail and the Possible Role of Photovoltaics

In order to exploit RES capacities for electricity generation successfully, it is of prime importance to mobilize *all the RES technologies* in a way coordinating time and economy. Contributions of importance for energy supply can in the short term (i.e. by 2010) only be made by technologies which are near to competitiveness and already supply considerable shares at present. Apart from hydro power, this especially applies to wind energy and biomass on both national and global scales. There is also the necessity of introducing all the other RES technologies into the market to a degree that enables their contribution to cover the future energy supply after 2010 with relevant shares. In the case of geothermal and solar thermal electricity production, it is therefore necessary to build a sufficiently large number of power plants within the next decade and to assure their contribution to the worldwide power plant market from 2010 on at the latest. The market for photovoltaics has to increase in such a way that after 2010 a share of 1 % of the electricity production can be achieved without further delay. Only then can a sufficiently quick exploitation of the potential be secured in order to maintain the above growth rates of RES for several decades.

Coordinated use of RES is also necessary to be able to guarantee the availability of RES electricity with high shares in electricity supply in as economical a way as possible, and to be able to keep the capacity for reserve and power balance as low as possible. Unusable surpluses of wind and solar electricity and extensive measures for storage should also be avoided, which can be managed by combining all the RES technologies with their quite different characteristics and by linking numerous plants as extensively as possible, e.g. within a European electricity network. This network should include extensive decentralized feeds as well as electricity transportation from areas with big and very steady supply (e.g. off-shore wind power, hydro power from Scandinavia; solar electricity from South Europe/North Africa) [TAB 2000]. Taking into account these criteria and the considerations of other global scenarios, the dissemination of RES in the above global scenario can adopt the structure shown in Fig. 1.11. Until 2020, further dissemination of hydro power follows along the considerations of the IEA [IEA 1998]; afterwards it will slow by reason of its limited potential. All in all, its share will have doubled by 2050

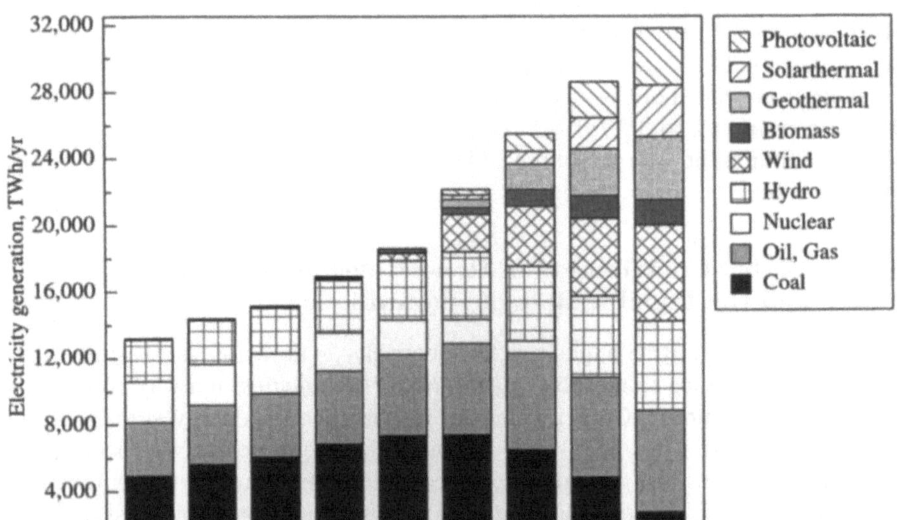

Fig. 1.11. Growth of renewable energy technologies in the "Solar Energy Economy" scenario until 2050

compared with its present 2 635 TWh/yr. Hydro power will therefore remain the most important renewable energy source for electricity production until almost the middle of this century and will then supply 17 % of the global electricity produced (presently 18.3 %).

Wind power is an energy technology which is developing in an extremely dynamic way, with worldwide growth rates of around 3 000 MW/yr at present. In scenario, its increase is oriented towards achieving a 10 % share of the global electricity supply in 2020, according to the European Wind Energy Association [Wind 1999]. Including offshore plants, there is presently a capacity of 900 GW installed, which is an average growth rate of 20 %/yr. If this growth dynamic is continued, wind power will have caught up with hydro power by 2040 at decreasing growth rates, and will exceed it in 2050 with a share of 5 800 TWh/yr or 2 200 GW. Wind power will then contribute 18 % to global electricity production.

A further intensified establishment of biomass and geothermal energy as energetically relevant technologies for electricity generation requires growth rates of about 10 %/yr until 2010, which will be achieved quite easily with regard to their favorable economic data and their technological maturity if appropriate priorities are established in energy politics. In the case of biomass, these priorities include, on the one hand, burning it in existing coal power plants, and on the other, intensified building of decentralized cogeneration plants mostly on the basis of wood and biogas. Apart from further tapping favorable geological resources (particularly USA; in Europe: Italy and Iceland), the dissemination of geothermal energy calls for rapidly entering into electricity generation based on hot-dry-rock (HDR) technology. Electricity generation with biomass and geothermal energy is extremely important for effective development of the electricity market with big shares of RES due to its guaranteed availability, its possible high annual availability (biomass in cogeneration plants about 4 000 h/yr, geothermal energy about 7 000 h/yr) and its low costs. Concerning biomass, a decrease in growth is assumed after 2030 because of its limited potential, which will lead to a maximum share in electricity supply of 1 500 TWh/yr (almost 5 %) or 375 GW in 2050. In contrast to this development of biomass, in the case of geothermal energy a further increase can be assumed due to the considerable large potential of HDR electricity production. Once established, geothermal energy with 3 800 TWh/yr (600 GW) in 2050 will play an important part in an electricity supply based on RES.

Solar irradiation merely contributes to the global energy supply with approximately 2 TWh/yr at present. In the long term, however, it will have to become the main contributor to a global renewable energy supply because of its virtually "unlimited" potential. The technologies for using solar irradiation therefore have to be developed with extreme intensity to enable them to play an adequate role among RES within the next few decades. Both technologies, i.e. solar thermal and photovoltaic electricity generation, are necessary, should large-scale substitution of the required fossil energy generation be successfil, in view of climate protection and resources, with the high growth rates in electricity consumption in the developing countries. Points which favor solar thermal electricity production are (solar) electricity generation costs of less than 10 cents/kWh, which are very favorable at present, along with the possibilities of combining it with existing power plants by means of hybrid operation and incorporating it into grid structures. In the longer term the guaranteed availability of (thermally stored) electricity is very important. Photovoltaics are favored by their enormous flexibility with respect to size and fields of application, by their existing competitive partial market, i.e. especially off-grid electricity supply, as well as by their robustness, long life-time, and low maintenance required. In the longer term, the remarkable potentials for cost reduction due to increasing markets and technological progress will be of growing importance.

The two technologies thus complement one another in a very favorable way in the necessary exploitation of the potentials of solar irradiation [TAB 2000]. Therefore comparable growth rates are assumed in the scenario for the two technologies. For solar thermal power plants there are concrete concepts for further development until 2010 [Trieb 2000] which are used as a basis and claim that with average growth rates of the market by 25 %/yr a share of 1 % will be exceeded in 2020 with 250 TWh/yr (or 70 GW). Further continuing growth of approximately 10 %/yr will lead to a contribution of this technology of 3 100 TWh/yr or 600 GW in 2050. Solar thermal power plants will then contribute 10 % to the global electricity supply.

Comparable goals, i.e. a share exceeding 1 % in 2020 and 10 % in 2050, can be fixed in the framework of this scenario for the market development of photovoltaics, which requires an average global market growth of 30 %/yr until 2010, 15 %/yr from 2010 to 2020, and a further 4 %/yr on a high level in the following 30 years. In this way, the share of PV in the global electricity supply will reach the data given in Table 1.6 (and Fig. 1.12, upper curve).

Compared with the period of 1993 to 1998 with an average market growth of 20 %/yr, the goal fixed in the scenario requires a considerable increase in the PV market growth rate in the next decade. Such an increase with an annual turnover of 6 500 MW_p/yr in 2010 could result in 40 times the present global market volume. Growth rates of this kind are quite likely to be reached. For example, in the last decade the market for wind power in Germany has increased by 30 %, and the cumulative capacity by an average 48 %/yr. The global PV market, too, has already increased by 30 %/yr since 1996. Nevertheless, these assumptions may be an optimistic upper limit on the future market development of photovoltaics, according to the ambitious goals fixed in the scenario described.[5] Downwards, the future market development of

[5] In scenarios A1 and C1 of the WEC [WEC 1998] a contribution of about 4 500 TWh/yr in 2050 has been assumed (1 500 TWh/yr in the scenario B), but without considering electricity generation by solar thermal power plants.

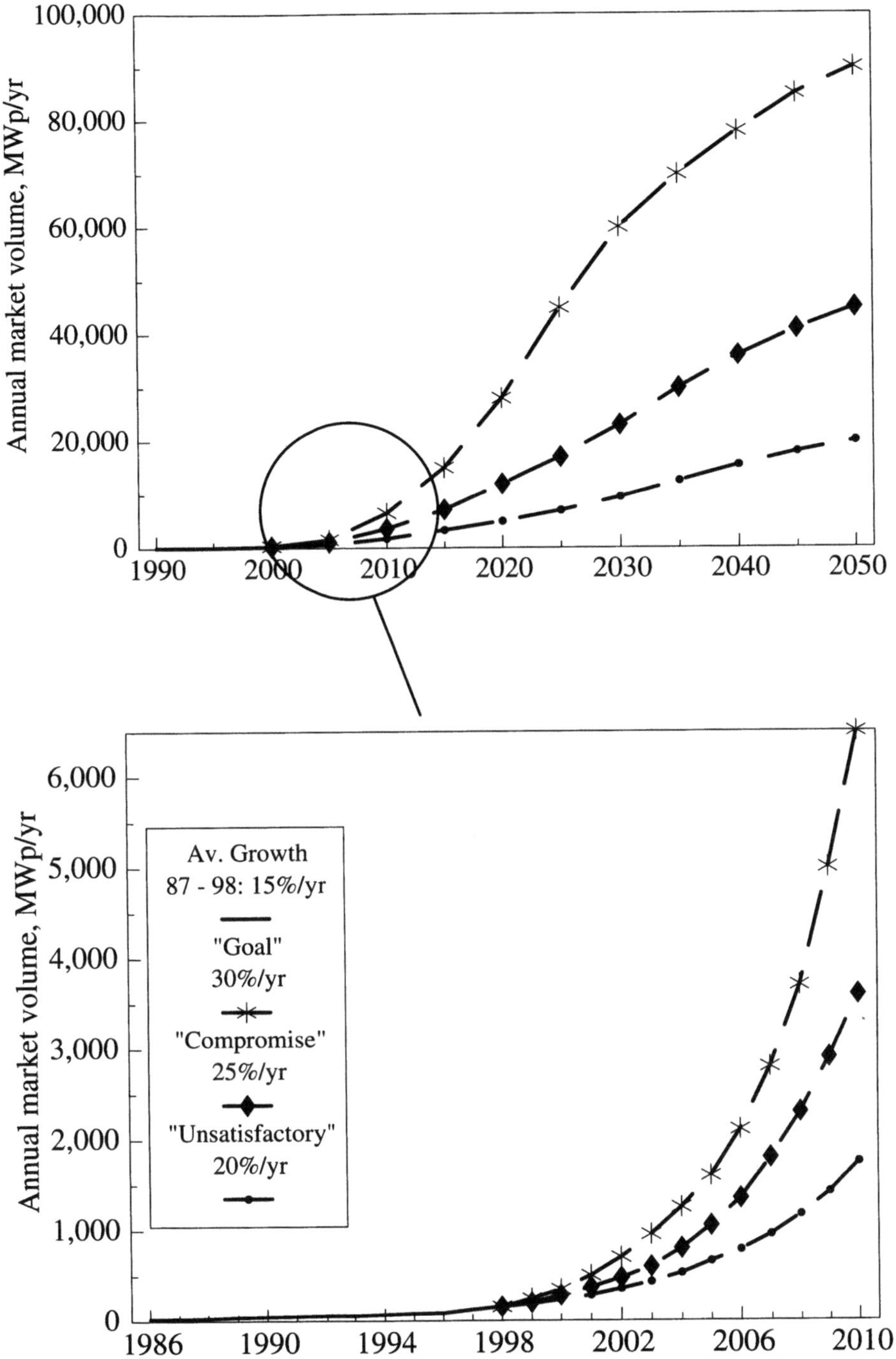

Fig. 1.12. Bandwidth of future development of the global annual PV market volume until 2050 (above) and in the next decade (below)

Table 1.6. Development of installed capacity, electricity generation and annual market volume of photovoltaics since 1987 and within the "Solar Energy Economy" scenario

	Installed capacity, MW_p	Electricity generation, TWh/yr	Share of total consumption, %	Market volume, MW_p/yr	Average market growth, %/yr
1987	128	0.15	0	28	
1993	400	0.50	0.003	61	13
1998	950	1.10	0.007	158	20
2010	26 650	37	0.2	6 500	30
2020	242 000	337	1.5	28 000	15
2030	767 000	1 078	4.2	60 000	7.5
2040	1 507 000	2 135	7.5	78 000	2.5
2050	2 381 000	3 400	10.7	90 000	1.4

photovoltaics can be limited by the growth rate necessary to stimulate the dynamics of development, the commercial basis for new and bigger production sites and adequate development of the partial markets. The lower limit can be fixed at an annual market growth of 20 %/yr, which is 5 percentage points more than the average value of the last decade. With an annual turnover of 1 750 MW_p/yr in 2010, such growth will lead to a market 10 times the present one (Fig. 1.12), and to a cumulative capacity of 9 500 MW_p. Analogous continuation of this growth development would not, however, lead to exceeding a 1 % share in worldwide electricity production until 2035, which would be too late to influence the energy supply structure relevantly and in time. Thus, in any case, stronger growth between 25 %/yr and the above 30 % is to be aspired to in the next decade, as indicated in Fig. 1.12 by the two upper growth curves.

1.3.3 Prospects for Renewable Energies in Europe and the Contribution of Photovoltaics

The "Energy Outlook to 2020" of the European Commission's Directorate General for Energy (DGXVII) dated November 1999 is a topical study on the possible energy supply of the 15 EU member countries [EU Outlook 1999] which also deals with the potential role of RES, among other things. The change in future energy supply given there is characterized as a "business as usual" development, i.e. existing trends will continue and structural changes will generally only be allowed within the framework of the economic parameters stated in the model. Thus this continuation will, among other things, also lead to an increase in CO_2 emissions, which is in opposition to the goals fixed at Kyoto. The development of RES will also take place quite slowly, based on the level for energy prices assumed even within the framework of the maximum version (sensitivity scenario S6). According to the basic version, in 2020 RES will reach a share in primary energy consumption of a mere 6.2 % (1195: 5.3 %), and 8.4 % in scenario S6.

The recommendations of the EU Commission's "White Paper" [Weissbuch 1997], on the other hand, claim that there will be a doubling of the RES share to approximately 12 % by 2010. In the "Campaign for Take-off" [Kampagne 1999] this goal of doubling the RES share is described in more detail. One goal is, for example, the dissemination of photovoltaics from about 200 MW_p at present to 3 000 MW_p in 2010 or 1 000 MW_p by 2003. The "partial goals" for the European Union derived from the global goals discussed in Sect. 1.3.2 also call for considerably greater efforts for the dissemination of RES than assumed in the topical EU Energy Outlook. For the field of electricity generation, these partial goals are compared in the following to the data given in [EU Outlook 1999].

A further strong increase in electricity consumption is a fundamental characteristic of the EU business-as-usual development (Fig. 1.13, upper curve). In this respect it is quite similar to the global scenarios of the IEA (see Fig. 1.9). It is therefore understood that efforts for effective substitution of fossil sources are much more difficult. As a result, the share of RES in the electricity supply of EU 15 will have risen from the present 14 % to only

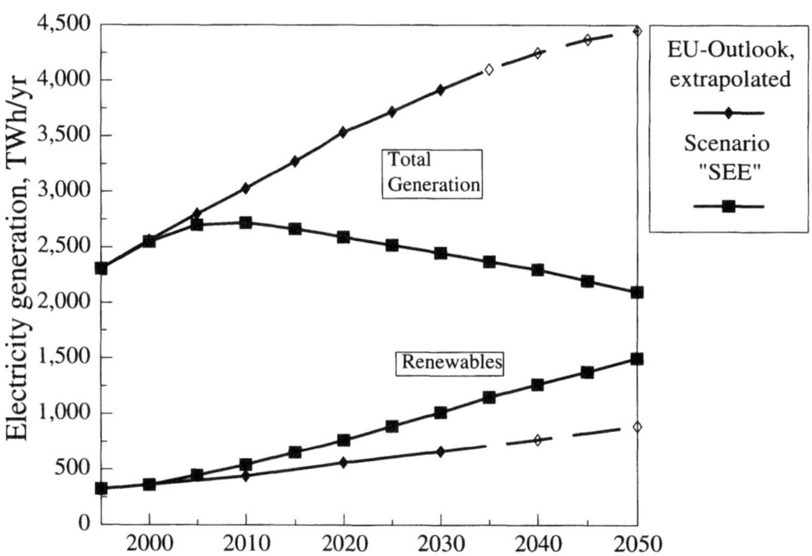

Fig. 1.13. Development of future electricity generation and the share of renewable in EU 15 corresponding to the EU Outlook [EU Outlook 1999] and the "Solar Energy Economy" scenario [Nitsch 1999]

17 % in 2030, and with extrapolation of the data from [EU Outlook 1999], to 20 % by 2050. The absolute difference between the total electricity demand and the RES share is steadily increasing, although there is relatively strong dissemination of wind power and biomass in particular, which will lead to a doubling of the RES share by 2030 (Fig. 1.13, lower curve). Considering the protection of climate and resources, simply disseminating RES without at the same time making more efforts to achieve considerably more rational energy (or in this case electricity) use is therefore not enough.

However, if the assumptions on more efficient use of electricity for the OECD countries made in the global scenario "Solar Energy Economy (SEE)" are considered (Fig. 1.10), the total electricity consumption of EU 15 will decrease after 2010 and will be only 65 % of the value given in [EU Outlook 1999] in 2030, but which is still the present amount of consumption. As a consequence, the effectiveness of a replacing fossil energy with RES is much greater in this scenario. At the same time, the goals for climate protection as well as the problems of availability of resources and risk minimization require considerably stronger mobilization of RES. As a deduction from the global approach of the SEE scenario [Nitsch 1999] and in accordance with the assumptions of the EU's White Paper, RES in EU 15 is to be developed as shown in Fig. 1.14 for electricity generation, as an example. Accordingly, the contribution of RES will increase from the present 325 TWh/ar (1997) to 540 TWh/yr in 2010 and to 765 TWh/yr in 2030. In the long term it will be 1 500 TWh/yr (2050). In 2035 wind power with 130 GW [Wind 1999] will take the place of hydro power as the most

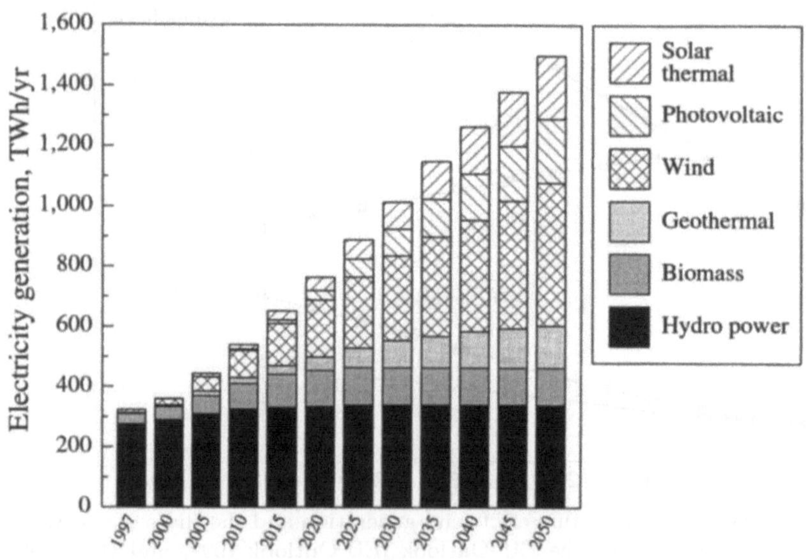

Fig. 1.14. Renewable electricity generation of EU 15 by shares of energy sources

important RES source, but all other RES will also contribute substantially to electricity supply in Europe from 2020 onwards. In connection with a more rational use of electricity, RES shares in the electricity supply will rise significantly, to 20 % in 2010, 30 % in 2020, and 42 % in 2030, and will hold from 2040 onwards the biggest share in the European electricity supply (55 %).

The market development of photovoltaics, which fits into this RES structure, complies with the goals already set for the global market. Their share should exceed 1 % in 2020, and by the middle of this century 10 % should be reached (Table 1.7). Only in this way can photovoltaics meet expectations as an important energy technology of the next century. The first goal is, in accordance with the dissemination plans of the White Paper, to have 3 000 MW_p in 2010, which means a share of 0.13 % in electricity generation. Thus by that time the market volume has to be 20 times the present one, which means a mean growth rate of 25 %/yr. During the last decade the European market has increased by an average 15 %/yr. If such market dynamics are actually reached, further development until 2050 with correspondingly decreasing growth rates of the market will be plausible and the share of 10 % (in the reduced electricity consumption of the SEE scenario!) will probably be achieved.

Apart from this upper bound on market growth to be aspired to, there is also a lower one, the value of which can be estimated according to the overall development of RES in the trend scenario of [EU Outlook 1999]. There, the

Table 1.7. Development of installed capacity, electricity generation, and annual market volume of photovoltaics of EU 15 since 1992, and optimistic development until 2050

	Installed capacity, MW_p	Electricity generation, TWh/yr	Share of total consumption, %	Market volume, MW_p/yr	Average market growth, %/yr
1992	30	0	0	6	
1998	115	0.1	0	21	20
2005	865	1.0	0.04	200	32
2010	3 000	3.7	0.13	650	24
2015	11 250	13.7	0.5	1 650	18
2020	26 250	32.3	1.2	3 000	12
2025	46 250	57.1	2.3	4 000	6
2030	69 300	85.9	3.5	4 600	3
2040	119 000	149.0	6.5	5 100	1
2050	171 000	215.0	10.2	5 300	0.4

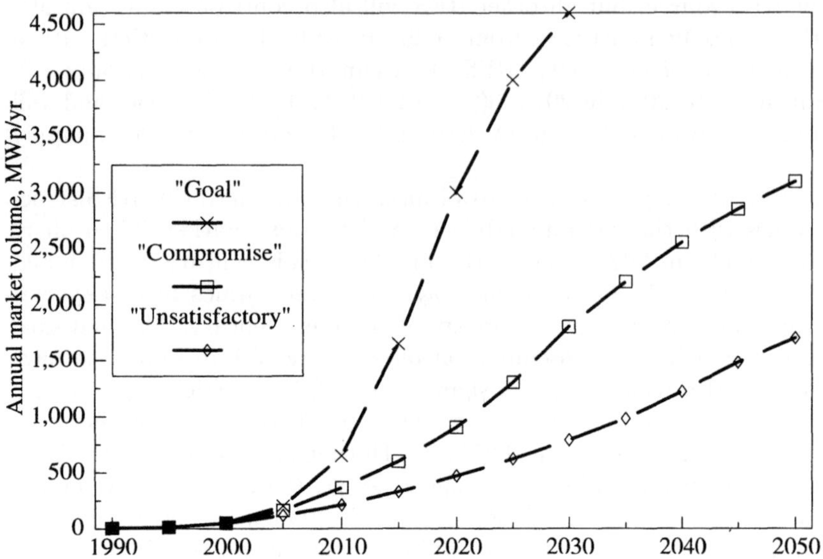

Fig. 1.15. Bandwidth of future development of the European (EU 15) photovoltaic market volume (upper line "Goal" see also Table 1.7)

present market growth at an average 20 %/yr will remain constant until 2010 (Fig. 1.15), which would lead to a market volume of a mere 200 MW_p/yr and a cumulative capacity of 1 400 MW_p in 2010. It is obvious that such development would allow relevant contributions of photovoltaics (> 1 %) to future energy supply only from 2040 onwards, which would hardly justify their present supporting incentives and investments. Such slow market growth must therefore be considered unsatisfactory. "Moderate" market growth, which would be required at least with initial growth rates of 24 %/yr, would result in a market with 365 MW_p/yr in 2010 and a cumulative capacity of 2 100 MW_p. In this case a 1 % share could possibly be reached in 2030.

Table 1.8 finally shows how these scenarios relate to the now numerous goals and market considerations of other studies, taking the installed capacity in 2010 as an example. Apart from the very theoretical considerations in the "Best practice" scenario of TERES II, the estimates range between 700 and 3 000 MW_p of installed capacity in 2010. Following the considerations in this paper, the contributions of photovoltaics to the European electricity supply in 2010 will be located in the upper third of these estimations.

Table 1.8. Different PV objectives set at the European level for 2010 and comparison with the upper scenario

Report/Investigation	Installed capacity in 2010, MW_p
Installed capacity end of 1998	115
ALTENER 1 (1993); extrapolated from 2005	1 200
TERES I (1993)	1 135
PV 2010; "Business as usual" scenario (1995)	1 200
PV 2010; "Strategic action plan" (1995)	2 000
TERES II; "ExternE Intern." scenario (1997)	700
TERES II; "Best Practice" scenario (1997)	12 000
White Paper (1997)	3 000
This paper: "Unsatisfactory"	1 400
This paper: "Compromise"	2 100
This paper: "Goal"	3 000

1.4 Prospects for Renewable Energies in Germany and the Contribution of Photovoltaics

An overview of the most topical scenarios in Germany shows the estimates for future dissemination of RES in electricity supply by 2010 differentiated between trend scenarios and (goal-oriented) dissemination scenarios (Fig. 1.16, [Prognos 1995, Prognos 1998, BMU 1999, Enquête 1995] [Altner 1995, Wuppertal 1998, Prognos 1999, Politik 1999]). There are more favorable values in all the more recent scenarios, which is the result of the surprisingly high dynamics of the dissemination of wind power. Also the trend scenarios with higher RES shares will probably be surpassed by 2010, as suggested by the share of RES already in electricity supply given for the end of 1999, with 5.4 %. The goal of doubling shares compared with 1997, designed into most of the dissemination scenarios with more than a 10 % contribution, will probably be achievable thanks to the rapid dissemination of wind power in the field of electricity supply.

After having reached the goal of doubling shares in 2010 with exceeding 10 %, the dissemination scenarios assume a further considerable increase in the REs share in electricity supply. Concerning electricity supply, this means that after 10 % is exceeded in 2010, the same is expected to happen for 20 % within a further decade (Fig. 1.17). Only the Prognos scenario of 1998 expects growth dynamics to flatten, so that merely 15 % will be exceeded.

For the longer term the scenarios assume a continuous increase in RES, which will result in their almost reaching or even exceeding 30 % in 2030, and probably exceeding at least 50 % by the middle of the next

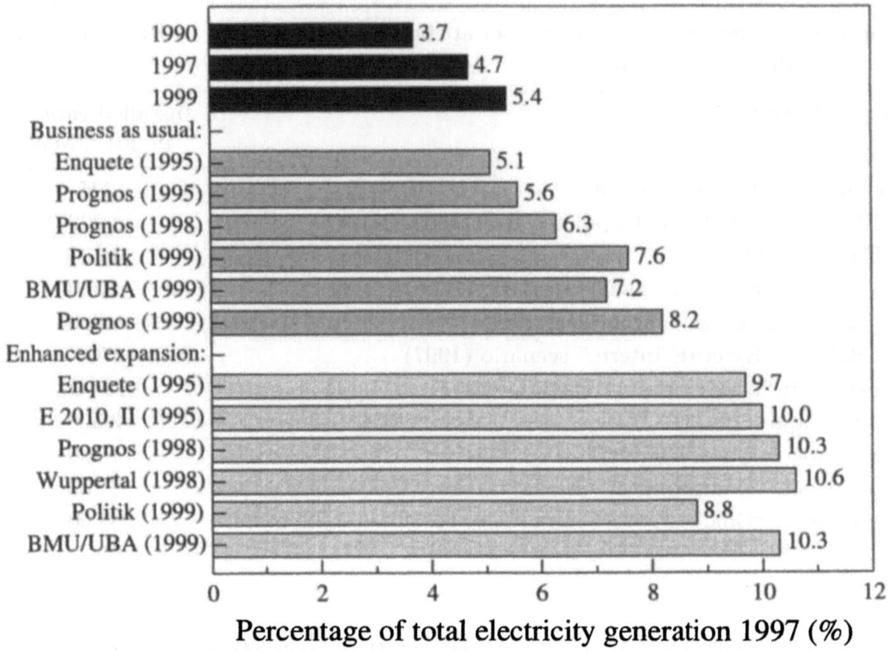

Fig. 1.16. Shares of RES in 2010 of total German electricity generation for 1997 for different scenarios

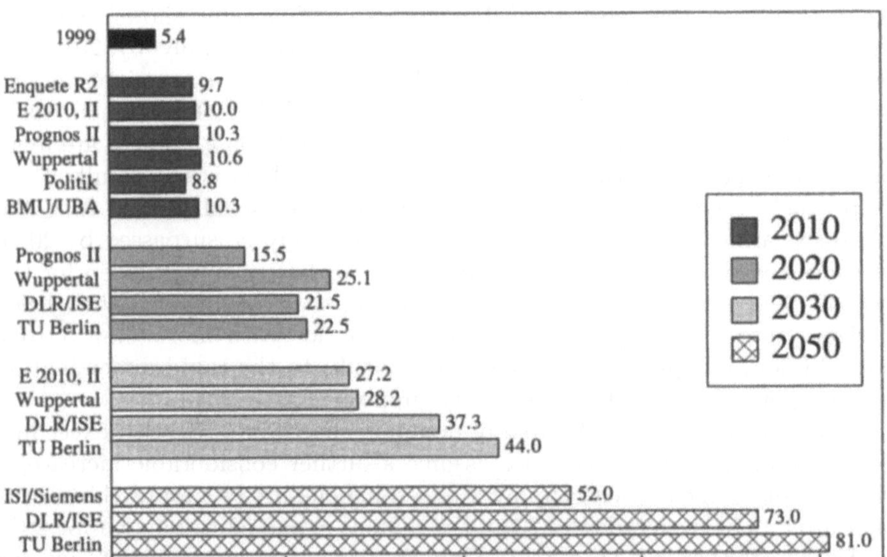

Fig. 1.17. Shares of RES in total German electricity generation between 2010 and 2050 for different scenarios

century.[6] The ISI/Siemens model calculations [ISI 1997] show that with a projected 80 % reduction in CO_2 emissions and without nuclear energy at the same time, RES shares will have to reach this value by 2050. The TU Berlin scenario [Quaschning 1999] assumes the highest RES shares with 44 % (2030) and 81 % (2050) exceeding the DLR/ISE scenario [Nitsch 1997]. According to the given goal-orientated scenarios, RES can thus become the dominat energy source by the middle of the next century, which shows that from the point of view of potential and structure such dissemination is considered possible. However, the tools and measures suggested in various scenarios for achieving these goals have been discussed in detail only up to 2010. They point out that in addition to the degree of present support, further support of energy politics is necessary in order to reach even this partial goal. Thus it cannot be taken for granted that more ambitious goals for the longer term will be put into practice. The targeted RES shares, however, point to the degree of the still required changes in the electricity (or energy) supply system, the related investments and points where investments have to be made, as well as to the resulting technical and economic consequences.

The scenario analysis allows the development of a "representative scenario" for the dissemination of RES in the case of goal-oriented and continuous energy politics towards developing RES potentials in an optimum way. Figure 1.18 shows an example of such a representative scenario [TAB 2000] which is compatible with the global and European development described above for a "Solar energy economy" (SEE scenario).

It is essential for systematic future development of the capacities for RES electricity generation to have the development co ordinated for all potentials with respect to time and economy. In the short term (i.e. until 2010) essential criteria are technical maturity and a status of already being close to economic competitiveness, which is in favor of biomass, wind power, and the remaining potentials of hydro power. Restrictions concerning potential are not yet expected except for hydro power, and in some regional coastal areas for wind power. Additionally, it is necessary to introduce all the other RES technologies into the market at least to a degree that enables them to participate in covering the future electricity demand to a relevant degree after 2010. This applies to photovoltaics, geothermal electricity generation (building HDR demonstration plants), and building a certain number of solar thermal power plants for local electricity supply in countries with high irradiation. There are no technological difficulties at all in contributing a share of about 10 % to covering the electricity demand of 2010 (half of which is in fluctuating form). Only some measures to reinforce the grid for feeding in wind power will be necessary.

[6] In all the scenarios electricity consumption still increases, if only slightly in some of them. The real shares of RES electricity are therefore lower in the scenarios than they are in the corresponding charts above, which means for the DLR/ISE scenarios 32 % in 2030 and 63 % in 2050.

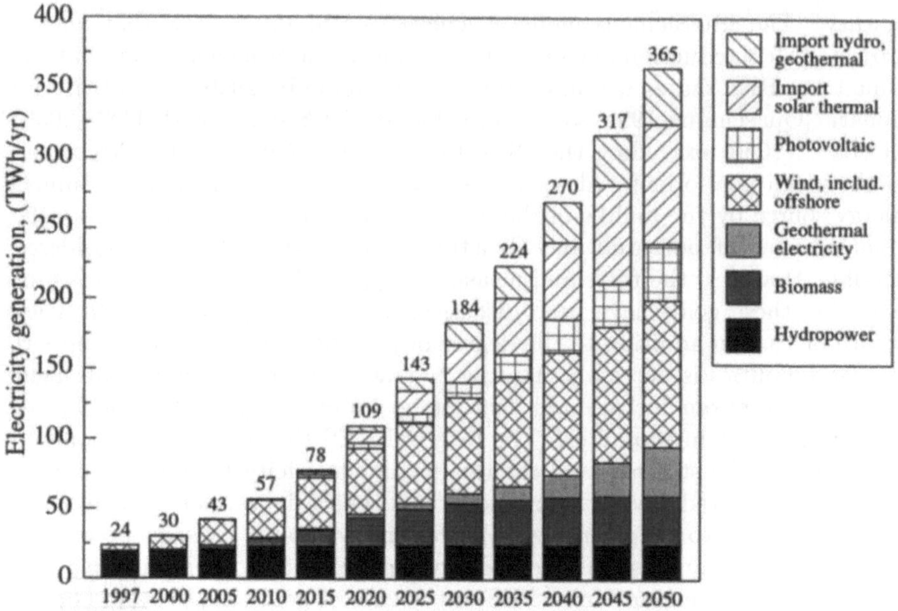

Fig. 1.18. A representative scenario for future RES development in Germany assuming a favorable energy policy and considering a suitable division among the different RES technologies

According to the representative scenario all RES technologies will play a part in the market after 2010 with increasing growth rates, possibly leading to a market share of around 20 % by 2020 and 35 % by 2030. Fluctuating and non-fluctuating shares will increase in approximately the same way. With shares of 10–15 % of electricity from wind and photovoltaics, there will not be any exorbitant demands made on fossil power plants until 2030, which would not have been considered anyway with the new investments required until that time anyway. For starting to import electricity, it is necessary to set up high-voltage direct current lines to South Europe/North Africa and to have them expanded in the Scandinavian region by 2020. After 2020 the quantities of imported electricity will increase considerably. Under the conditions mentioned above, shares of total RES electricity can be expected to exceed 50 % (compared to present electricity generation) approximately in 2040, and to reach 75 % by the middle of the century. With respect to their potential, electricity generation with hydro power (by 2020) and biomass will be developed entirely, while the other domestic technical potentials will only be exploited by approximately 30 to 35 %. Import potentials will still be available to a large degree. Thus, from the point of view of technical potential, there will be margins great enough to cover the electricity demand with RES almost entirely, if a strategy for balanced development of all RES sources or the respective technologies is pursued.

The shares of individual kinds of energy sources range between 5 and 15 % or reach a maximum 20 % at massive exploitation (e.g. wind). Therefore they complement each other in their generation characteristics. The share of clearly fluctuating RES (wind, photovoltaics) in the total electricity production is 30 % (line 1 in Table 1.9, [TAB 2000]), for imported electricity it is 25 % (line 8). Due to the possibilities of having it stored in solar thermal power plants, imported electricity is not liable to short-term fluctuations. With hydro power there are only seasonal fluctuations. Electricity generated with geothermal power plants is equal to a fossil base load supply. Additionally there are also domestic RES sources with little fluctuation that accounts for almost 20 % (line 4). Approximately the same RES shares are supplied from sources in North Germany (line 15) and South Germany (line 16) into the German grid, as the imported quantities are fed in either in the south or the north according to their origin. Thus there will not be an unusual, permanent electricity transport from north to south within Germany, but rather an alternating short-term balancing of capacity in both directions. With this mixture of RES technologies a guaranteed electricity supply without any solar surplus and extraordinary strain of the grid can be secured with RES shares of even 75 % in total electricity generation, if there is an appropriate adaptation of

Table 1.9. Structure of German electricity generation in 2050 with about 75 % share of renewables (related to net electricity generation of 1997 = 509 TWh/yr) derived from a representative scenario (see Fig. 1.18, [TAB 2000])

Origin	No.	Technology	Energy, Twh/yr	Capacity, GW	Full load hours, h/yr
Inland	1	Fluctuating RES	145	87	1 665
	2	Photovoltaics	40	45	900
	3	Wind including off shore	105	42	2 500
	4	Minor fluctuating RES	95	20	4 800
	5	Hydro power	25	5	7 000
	6	Biomass	35	5	7 000
	7	Geothermal plants	35	5	7 000
Abroad	8	Minor fluctuating RES	125	22	5 700
	9	Solar thermal plants	85	16	5 500
	10	Hydro power/geothermal	40	6	6 700
Total	11	Domestic (1 + 4)	240	107	2 240
	12	Import (8)	125	22	5 700
	13	Fluctuating (1)	145	87	1 665
	14	Minor fluctuating (4 + 8)	220	42	5 240
	15	North Germany (3 + 10)	145	48	3 020
	16	South Germany (5 + 7 + 9)	145	26	5 580
	17	Distributed (2 + 6)	75	55	1 360
	18	Total RES (1 + 4 + 8)	365	129	2 830

the remaining (fossil) power plants (e.g. with quickly adjustable combined cycle power plants) and the consumers' electricity demand (e.g. structuring of tariffs according to solar supply).

It is understood that there are also quite different estimates concerning the future development of photovoltaics in the German market, similar to those for the European marked, ranging from conventionally modest estimates [Prognos 1995], [Prognos 1998], [Prognos 1999] to optimistic developments by DLR/ISE [Nitsch 1997] and TU Berlin [Quaschning 1999], which assume very favorable framework conditions fixed by energy politics. Figure 1.19 shows the evaluation of the development described above in detail [TAB 2000] until 2020. There, the topical favorable framework conditions for photovoltaics are already considered, like e.g. the 100 000 roofs programme and the modification of the electricity feed law with 99 Pf/kWh for photovoltaics. Apart from that, it has to be considered that the German production capacity will probably be at least 70 MW_p by 2004, a big share of which will have to be sold in the domestic market at least initially. Under this condition it should be quite possible to exceed 0.1 % by 2010 and to approach a 1 % share by 2020.

The market growth connected with this development and the cumulative capacity reveal (Table 1.10, Fig. 1.20) that it is only necessary for this pur-

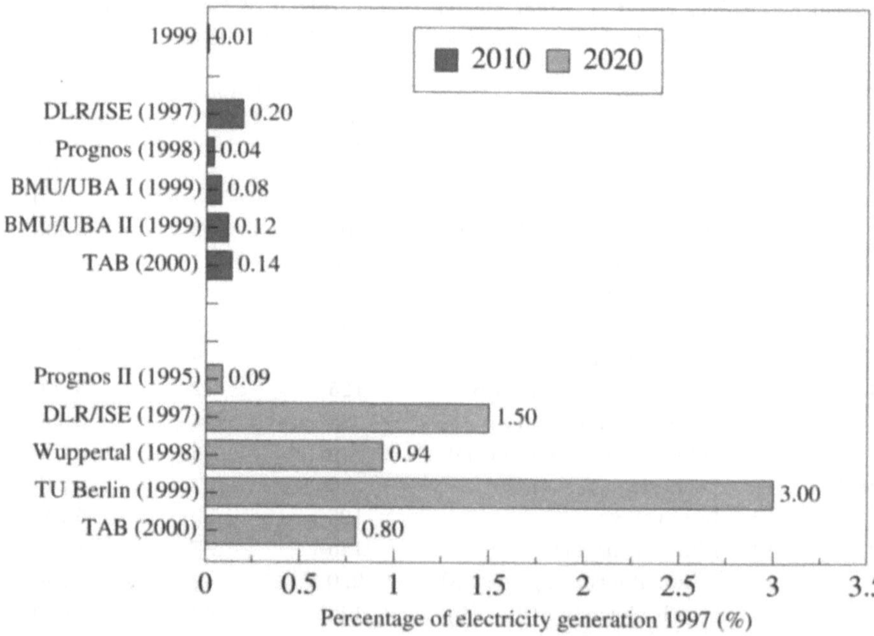

Fig. 1.19. Photovoltaics as a percentage of total electricity generation in 2010 and 2020 for different scenarios, including the value of the representative scenario of Fig. 1.18

Table 1.10. Development of installed capacity, electricity generation, and annual market volume of photovoltaics in Germany since 1992, and development until 2050

	Installed capacity, MW_p	Electricity generation, TWh/yr	Percentage of total consumption	Market volume, MW_p/yr	Average market growth, %/yr
1992	6	0.005	0	2.5	
1999	70	0.06	1.01	20	30
2005	375	0.33	0.07	80	23
2010	980	0.87	0.17	160	14
2015	—	—	—	—	10
2020	4 430	4.0	0.79	365	8
2025	7 370	7.0	1.38	530	7.5
2030	12 200	11.1	2.18	750	7
2040	25 500	34.4	4.60	1 350	6
2050	44 500	40.9	8.05	1 900	4

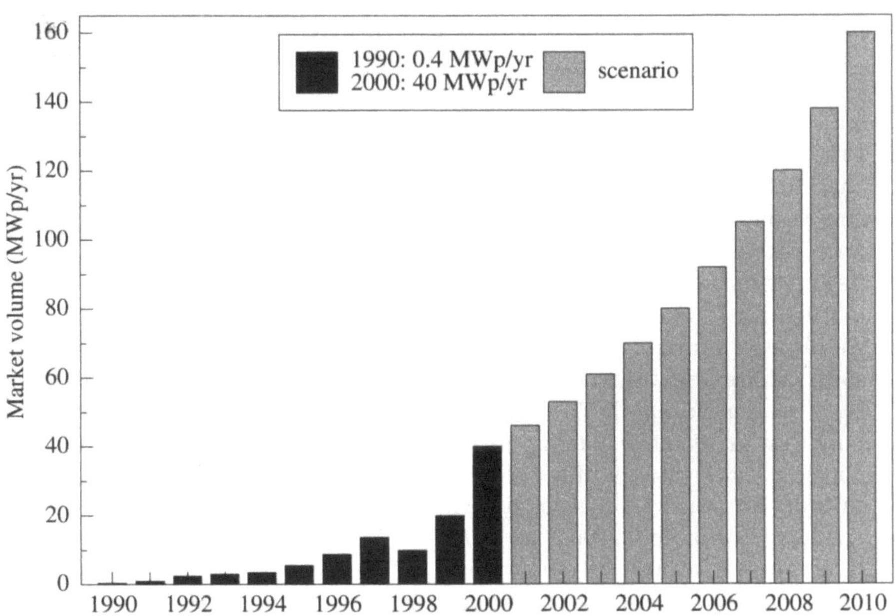

Fig. 1.20. Development of the German photovoltaic market according to the representative RES scenario for Germany

pose to keep up approximately the growth rate of the last years. In so doing, the installed capacity of almost 1 000 MW_p in 2010 corresponds to 14 times the value of 1999. With a slowdown in market growth there will be more than 4 400 MW installed by 2020, already securing a relevant share in electricity supply for photovoltaics. By the middle of this century the installed photovoltaic capacity can approach 50 GW in Germany, which would mean an integration of about 300 km^2 of module area in the settlement structure. Market volume would have grown to almost 2 000 MW/yr.

1.5 Conclusions and Recommendations

1. In this book the dissemination of RES for electricity generation follows global, European, and national scenarios which come up to the requirements for climate protection, and allow suitable growth in energy consumption for newly developed and developing countries, but at the same time aim at decreasing the share of fossil primary energies in the medium and long term, and reduce the share of nuclear energy towards zero. Developments like these would to a large extent meet the demands for sustainable energy supply. If all the RES technologies described are mobilized in a balanced way and in coordinated time sequence, i.e. according to their economic market relevance and their technical potential, the necessary dissemination will not be limited by low technical potential. Especially with photovoltaics, only about 5 % of the global technical potential will be used by 2050, even with optimistic growth rates.
2. Presently, the share of photovoltaics in electricity generation is clearly less than 0.01 % worldwide. In Germany this level has just been reached. To enable photovoltaics to play a part in shaping the future energy supply in an economically and energetically relevant way in the next few decades, its contribution must increase to at least 0.1 % by 2010 and 1 % by 2020. With a basis like that, the share of photovoltaics in global electricity supply will probably achieve a good 10 % in 2050, and thus a balanced sound share will be reached in the total RES contribution which could then be approximately 70 % according to the scenarios. It is important to note that in 2050 the demand for electricity will be about 30 000 TWh/yr, which is more than double the present consumption.
3. In order to reach the shares mentioned above for photovoltaics, average growth rates of the annual installations of PV systems between 25 %/yr (Germany) and 30 %/yr (worldwide, EU) are necessary. The market volumes to be deduced from these data for 2010 are 6 500 MW/yr worldwide, 650 MW/yr for the EU, and 150 MW/yr for Germany. Compared with the present installations (1998) of 160 MW/yr (worldwide), 20 MW/yr (EU), and 10 MW/yr (Germany), this implies an increase of 15 times (Germany) or 40 times (worldwide) the present amount. The cumulative power related to such a market expansion would increase to 26 500 MW worldwide, 3 000 MW in the EU, and 800 MW in Germany by 2010.

4. If the annual market growth remains lower than 20 %/yr in the near future (average of the last 10 years: 15 %/yr), a share of more than 1 % in electricity supply will probably be reached between 2030 and 2040 at the earliest, which will not be enough to significantly influence future energy supply in time. It is therefore of prime importance to stabilize the emerging acceleration of market growth (market growth from 1996 to 1998 by 30 %/yr) in the next decade or to expand it even further. For this purpose, effective instruments for market support in the electricity grid like e.g. the Act on Granting Priority to Renewable Energy Sources in Germany ("RES-Act") are crucial preconditions. They are required in similar form by all the industrialized countries to quickly increase the big market segment of grid-connected power plants and to enable considerable cost reductions. With the industrialized countries acting as pioneers, the financing of smaller PV plants and thus further dissemination in the field of off-grid supply in countries with less developed infrastructure would then be decisively facilitated.

References

[Altner 1995] Altner, G., Dürr, H.P., Michelsen, G., Nitsch, J., Zukünftige Energiepolitik - Vorrang für rationelle Energienutzung und regenerative Quellen. Economica-Verlag, Bonn 1995.

[BMU 1999] Nitsch, J., Fischedick, M. et al., Klimaschutz durch Nutzung erneuerbarer Energien. Study made in cooperation by DLR/WI/ZSW/IWR/Forum on behalf of the German Bundesministerium für Umwelt, Naturschutz und Reaktorsicherheit and the German Umweltbundesamt. Stuttgart, Berlin, November 1999.

[Cabraal 1999] Cabraal, A., Assuring quality in World Bank/GEF-funded solar home systems projects. Photovoltaic Global Approval Program Conference, Geneva, Switzerland, 3-4 May 1999.

[Cameron 1998] Cameron, M., Gostinelli, F., Helm, P., Bonda, J.C., Sharma, B.D., Removing Barriers Between Financial Institutions and the PV Market, Lessons from India. Proceedings of the 2nd World Conference on Photovoltaic Energy Conversion, Vienna, Austria, 6-10 July 1998, pp. 2950-2953.

[Enquête 1995] Energie und Klima. Final Report of the Enquête-Commission Schutz der Erdatmosphäre of the German Bundestag. Economica Verlag, Bonn 1995.

[EU Outlook 1999] Directorate General for Energy, European Union Energy Outlook to 2020. EU Commission, Brussels, November 1999.

[Giroult 1998] Giroult-Matlakowski, G., Theden, U., Blakers, A.W., Photovoltaic Market Analysis. Proceedings of the 2nd World Conference on Photovoltaic Energy Conversion, Vienna, Austria, 6-10 July 1998, pp. 3403-3406.

[Grubb 1993] Grubb, M., Meyer, N., Wind Energy Resources, Systems and Regional Strategies. Chapter 4 in [Johansson 1993] Island Press, Washington D.C. 1993.

[IEA 1998] International Energy Agency, World Energy Outlook. IEA/OECD, Paris 1998.

[IEA 1998] International Energy Agency, Photovoltaic Power Systems Program, Trends in PV Power Applications in selected IEA countries 1992-1998. International Energy Agency, December 1998.

[IEA 1999] Key World Energy Statistics, Int. Energy Agency, Paris 1999.

[ISI 1997] Toensing, E. et al., Energieszenarien mit reduzierten CO_2-Emissionen bis 2050. Energiewirtschaftliche Tagesfragen (42), Heft 8, 1997, p. 474.

[Johansson 1993] Johansson, T.B., Kelly, H. et al., Renewable Energy Sources for Fuels and Electricity. Island Press, Washington D.C. 1993.

[Kampagne 1999] EU Commission, Renewable Sources of Energy, "Campaign for Take-off". Service Paper DOCSEC(99)504, Brussels, April 1999.

[Klaiss 1992] Klaiss, H., Staiß, F. (eds.), Solarthermische Kraftwerke für den Mittelmeerraum. 2 volumes, Springer-Verlag, Heidelberg, Berlin 1992.

[Lovins 1999] Lovins, A., Hennicke, P., Voller Energie - Die globale Faktor Vier-Strategie für Klimaschutz und Atomausstieg. Campus Verlag, Frankfurt 1999.

[Nitsch 1997] Nitsch, J., Langniss, O., Luther, J., Wiemken, E., Strategien für eine nachhaltige Energieversorgung - Ein solares Langfristszenario für Deutschland. Forschungsverbund Sonnenenergie, Cologne, Dezember 1997.

[Nitsch 1999] Nitsch, J., Entwicklungsperspektiven erneuerbarer Energien und ihre Bedeutung für die Energieversorgung von Entwicklungsländern. Proceedings of the Conference "Märkte der Zukunft – Erneuerbare Energien für Entwicklungsländer" of the Wirtschaftsministerium Baden-Württemberg and the German Aerospace Center, Friedrichshafen, Germany, 17 November 1999, pp. 11–20.

[Politik 1999] DIW Berlin, FZ Jülich, FhG-ISI, Oeko-Institut, Politikszenarien für den Klimaschutz II. Szenarien und Massnahmen zur Minderung von CO_2-Emissionen in Deutschland bis 2020. Study on behalf of the German Umweltbundesamt, Berlin, June 1999.

[Posorski 1998] Posorski, R., Fahlenbeck, B., Technical Standards and Tender Specifications for Solar Home Systems and PV Supply of Rural Health Stations in Developing Countries. Proceedings of the 2nd World Conference on Photovoltaic Energy Conversion, Vienna, Austria, 6-10 July 1998, pp. 2859-2863.

[Prognos 1995] Prognos AG, Die Energiemärkte Deutschlands im zusammenwachsenden Europa - Perspektiven bis zum Jahr 2020. Basel 1995.

[Prognos 1998] Prognos AG, Möglichkeiten der Marktanreizförderung für erneuerbare Energien auf Bundesebene unter Berücksichtigung veränderter wirtschaftlicher Rahmenbedingungen. Study on behalf of the German Bundesministerium für Wirtschaft, Bonn, December 1998.

[Prognos 1999] Prognos AG, EWI, Die längerfristige Entwicklung der Energiemärkte im Zeichen von Wettbewerb und Umwelt. Study on behalf of the German Bundesministerium fuer Wirtschaft und Technologie, Berlin, Basel, November 1999.

[PV 2010 1995] Commission of the European Communities, Directorate General for Energy (ed.), European Photovoltaic Industry Association, Photovoltaics in 2010, vol. 3, The World PV Market to 2010. Brussels, September 1995.

[PV News 2000] PV News, vol. 19, No. 2, February 2000.

[PV Report] Curry, R. (ed.), Photovoltaic Insider's Report. Monthly newsletter.

[Quaschning 1999] Quaschning, V., Systemtechnik einer klimaverträglichen Elektrizitätsversorgung in Deutschland für das 21. Jahrhundert. Habilitationsschrift, TU Berlin 1999.

[Shell 1995] Energie im 21. Jahrhundert. Study of the Shell-AG Hamburg, Aktuelle Wirtschaftsanalysen 5, Heft 25 (1995).

[TAB 2000] Nitsch, J., Trieb, F., Potenziale und Perspektiven regenerativer Energieträger. Study on behalf of the German bureau für Technikfolgenabschätzung beim Deutschen Bundestag, Stuttgart, March 2000.

[Trieb 2000] Trieb, F., Competitive solar thermal power stations until 2010 - the challenge of market introduction. Renew. Energy 19 (2000), pp. 163-171.

[UN 1998] Bericht über die menschliche Entwicklung. UNDP-Bericht, Dt. Ges. für die Vereinten Nationen, Bonn 1998.

[WEC 1995] Global Energy Perspectives to 2050 and Beyond. Joint IIASA-World Energy Council Report. Luxembourg, London 1995.

[WEC 1998] Energie für Deutschland - Fakten, Perspektiven und Positionen im globalen Kontext. Deutsches Nationales Komitee DNK of the World Energy Council. Duesseldorf 1998.

[Weiss 2000] Weiss I., Sprau H. P., Die Entwicklung des Photovoltaik-Marktes in 1998 und seine Stellung im internationalen Vergleich. 15. Symposium Photovoltaische Solarenergie, Staffelstein März 2000.

[Weissbuch 1997] Erneuerbare Energien - Weissbuch für eine Gemeinschaftsstrategie und Aktionsplan. European Parliament, board for research, technological development and energy. COM (97)599 final (26/11/1997), Strasbourg 1997.

[Wind 1999] European Wind Assoc., Forum for Energy and Development, Greenpeace Int., Windstärke 10. Studie zum weltweiten Ausbau der Windenergie, Oktober 1999.

[Winter 1988] Winter, C.M., Nitsch, J., Hydrogen as an Energy Carrier - Technologies, Systems, Economy. Springer, New York 1988.

[World Atlas 1998] World Atlas and Industry Guide 1998-1999. Int. Journal on Hydropower & Dams, Sutton, UK 1998.

[Wuppertal 1998] Wuppertal Institut für Klima, Umwelt, Energie GmbH, Der Beitrag des Sektors Bauen und Wohnen für den Klimaschutz. Study on behalf of the German Ministerium fuer Bauen und Wohnen of the state of North Rhine-Westphalia, Wuppertal 1998.

2 Solar Modules and Photovoltaic Systems

J. Luther, K. Preiser, and G. Willeke

Solar modules are electronic devices that convert sunlight directly into electricity. The following chapter describes the working principles as well as the technology behind those devices, which are in industrial production today. Furthermore solar cell schemes are discussed that may play a role in the power market of the years to come.

2.1 Basic Principles of Photovoltaic Energy Conversion

The physical nature of light is difficult to understand. On the one hand light can be characterised as an electromagnetic wave like radio waves and x-rays with a characteristic frequency and wavelength. In this view sunlight consists of a spectrum of electromagnetic waves, i.e. a whole range of different wavelengths, including the ultraviolet and infrared wavelengths which are invisible to the human eye as well as of course all the visible colors. The wave-like properties of light are needed for example to explain optical effects like the blue color of a typical crystalline silicon solar cell. This color is due to the thin antireflection coating on the surface of the cell, a coating which is not perfect in the sense that it reflects blue light more than the other visible wavelengths. At the same time light may be considered to be composed of particles – so-called photons – that carry a characteristic energy and momentum. The short wavelength blue photons carry a higher energy than the long wavelength red photons. It is the particle-like behavior of light that is needed to understand the photovoltaic effect. Conversion of light into electricity in a solar cell is based on this effect.

A solar cell [Goe 1998, Gre 1986] is composed of a semiconducting material such as silicon which has electrical conduction properties somewhere between those of metals like silver or copper and insulators like glass and quartz. An electric current in such materials is carried by negatively charged elementary particles: the electrons. In semiconductors the energies of these electrons are restricted to energetically separated "energy bands." Important for the understanding of the photovoltaic effect are two bands: the lower lying so-called valence band and the energetically higher lying conduction band. The two bands are energetically separated by a bandgap. Under the condition of thermal equlibrium the valence band of a pure semiconductor is practically fully occupied by immobile electrons virtually "sitting" between neighboring semiconductor atoms. To the same extent the conduction band

Fig. 2.1. The internal photoelectric effect is based on photons of energy E_{ph} that excite electrons e^- from the valence band E_v into the conduction band $E > E_c$ across the bandgap E_g of a semiconductor. Photons having energies lower than Eg will not contribuite to photovoltaic electricity generation. The surplus kinetic energy of the excited electrons in the conduction band is converted to thermal energy, thus constititing a loss

is empty. Under these conditions no electrons can move about freely in the solid and the semiconductor behaves more or less like an electrical insulator.

There are several effects that can make a semiconductor electrically conductive. Two of them play a decisive role in a solar cell. The first effect is the so-called internal photoelectric effect illustrated in Fig. 2.1. Here a photon is absorbed in the semiconductor, thereby transferring energy to a bound immobile electron in the valence band which is then excited across the bandgap into the conduction band, where it is free to move about the solid. As we can see in the figure, the ultraviolet photon (PH_2) carries more energy than required to transfer the electron across the gap. The surplus energy is simply used to make the electron move about (represented by the skateboard); this process eventually leads to the generation of heat which is lost in the photovoltaic conversion process. The nfrared photon shown (Ph_1) does not carry enough energy to excite an electron across the gap. This infrared photon will therefore not be absorbed and the semiconductor is transparent to this long wavelength light.

What is not shown in Fig. 2.1 is the fact that the electron that has been excited to the conduction band leaves a place in the valence band that is now not occupied by an electron. Such an empty energy state of a negatively charged electron behaves like a positively charged particle – a so-called hole. These charge-carrying holes are also comparatively free to move about the semiconductor solid. Thus both electrons and holes produce an electric current under an applied voltage. Actually there is an attractive force between the negatively charged electron (in the conduction band) and the positively charged hole (in the valence band) resulting in a locally confined electron-hole pair. The attractive force creating this pair has to be overcome and the electron-hole pair literally split spatially in order to produce a useful electric current. If left alone, the excited electron will eventually recombine either with the original hole of the electron-hole pair or any other hole within reach to produce a photon or in a competing process just waste heat. It is therefore essential for the operation of a solar cell to separate the charges in the photogenerated electron-hole pairs as soon as possible and to collect the charge carriers at the external electrodes within their lifetime, i.e. before they recombine.

Recombination processes between electrons and holes occur with high probability at disturbed crystal sites. Thus in order to achieve high lifetimes of the energetically excited electrons, high quality semiconductor material is usually required, which should be free of defects like unwanted inpurities and crystal imperfections to an extremely high degree. Surfaces obviously represent areas of crystal imperfections. Strong charge carrier recombination at open surfaces can only be avoided if these areas are "passivated" properly. This is generally done by depositing suitable coatings on the surfaces (for silicon, e.g., silicon nitride).

The second possibility to manipulate the electrical conductivity of a semiconductor is by introducing well-controlled small amounts of so-called dopant atoms. In silicon for example the electrical conductivity may be increased by several orders of magnitude simply by replacing every one-millionth silicon atom by a phosphorus or a boron atom. The phosphorus atom has one electron more than the silicon atom. This extra electron is immediately separated from its donating host atom which in turn becomes positively charged – a charge that is now fixed in space and thus immobile; the thermally liberated electron enters the conduction band. This leads to the fact that we have more freely mobile electrons than holes in the semiconductor, which is then said to be n-doped. In contrast to this, the boron atom has one electron less than the silicon atom. The boron atom easily binds an electron from the valence band and produces a hole there, leading to p-type conduction. This selective p- and n-type conduction forms the basis of all semiconductor devices.

A cross section of a typical silicon solar cell is shown in Fig. 2.2. In this material the bulk of the energy converter is composed of p-type material. Only a front surface layer of the wafer has n-type conductivity. This n-region is produced by diffusing phosphorus into it at high temperatures. Following the nomenclature of semiconductor electronics the n-layer is called the emitter and the p-region the base.

When forming such a so-called pn-junction diode structure, electrons from the emitter diffuse instantaneously into the base, and holes from the p-region diffuse into the emitter. This is due to the fact that the emitter contains compared to the base a very high concentration of electrons, whereas the

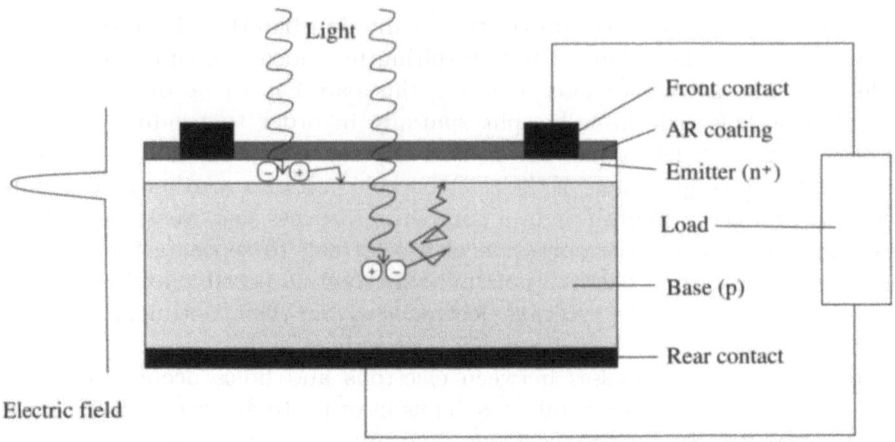

Fig. 2.2. Schematic cross section of a solar cell. Photogenerated carriers reach the electric field region of the junction between two different types of silicon semiconductor material: emitter n+ and basis p. Due to the assymetry in the silicon light absorber the negative charges are collected at the front contact while the positive charges move to the rear contact. AR = anti reflective coating

base is rich in holes. This diffusion of charge carriers leads to the build up of an electric field, and thus a resulting internal voltage in the vicinity of the pn-junction. Under equilibrium conditions the electrical forces due to this field compensate the forces driving the above mentioned diffusion – thus no net electric current flows.

If light noe impinges on the solar cell, part of it will penetrate and produce electron-hole pairs, provided the photons have sufficient energy. These added charge carriers result in a non-equilibrium condition in the cell. Under such conditions a superposition of the above mentioned electric and diffusion forces may separate the electron-hole pairs and transport them to the negative and positive electrode of the cell respectively. It is important that these transport processes can compete with the ever-present recombination opportunities of the energetically excited charge carriers.

The accumulation of charges at the external electrodes leads to the buildup of an external voltage between the metal contacts. This voltage may then drive the charge carriers through an external electrical load. The work delivered to this load is converted solar energy.

Under open-circuit conditions the maximum voltage, the open-circuit voltage V_{oc}, is measured between the contacts. V_{oc} increases with increasing bandgap of the semiconductor material. If we directly connect the front and rear metal electrodes electrically, we short circuit the voltage build-up and produce a short-circuit current I_{sc}. Due to the fact that under lower band-gap situations more solar photons are able to excite electron-hole pairs, the short-circuit current increases with decreasing band gap.

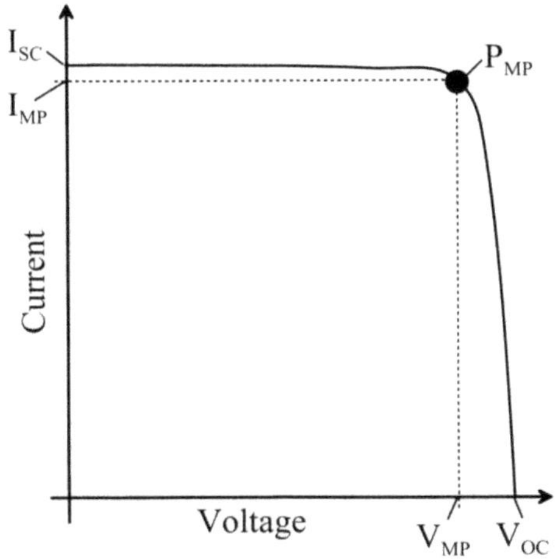

Fig. 2.3. Current-Voltage curve of a silicon solar cell. I_{sc} =short circuit current, V_{oc} =open circuit voltage, P_{mp} =maximum power point

Figure 2.3 shows a typical current-voltage curve of a solar cell in operation. The maximum power output of a solar cell, P_{mpp}, i.e. the maximum current-voltage product, is strongly related to the product of open-circuit voltage and short-circuit current. The two parameters depend in an opposite way on the bandgap energy. Given the terrestrial solar spectrum, an ideal semiconductor material having a bandgap of 1.4 eV should result in the greatest power output. The most important semiconductors for solar energy applications have bandgaps in the range 1.1–1.7 eV. (Photons of yellow light have an energy in the vicinty of 2 electron volts (eV); 1.4 eV is the energy of invisible "dark red" photons.)

The performance of solar cells is not only determined by material properties and semiconductor technologies but also by the optics of the cell. For example the fraction of light that enters the cell may be enhanced by putting an antireflection (AR) coating composed of a thin optically transparent dielectric layer on top of the solar cell. This leads to a "smoother" transition between the optical properties of the ambient air and the semiconductor material. A second way to reduce optical reflections at the surface of a solar cell is by geometrical texturing of the surface. Pyramidal structures are very effective for this purpose.

An optical technique that is essential for several thin-film solar cell concepts is radiation trapping. By means of multiple internal optical reflections, the light path of the photons in the semiconductor may be increased by more than an order of magnitude, thus leading to sufficient photon capture even in low-absorption materials like crystalline silicon.

In order to couple as much light as possible into the solar cell, the front metal contacts are formed as a narrow, widely spaced finger grid in present

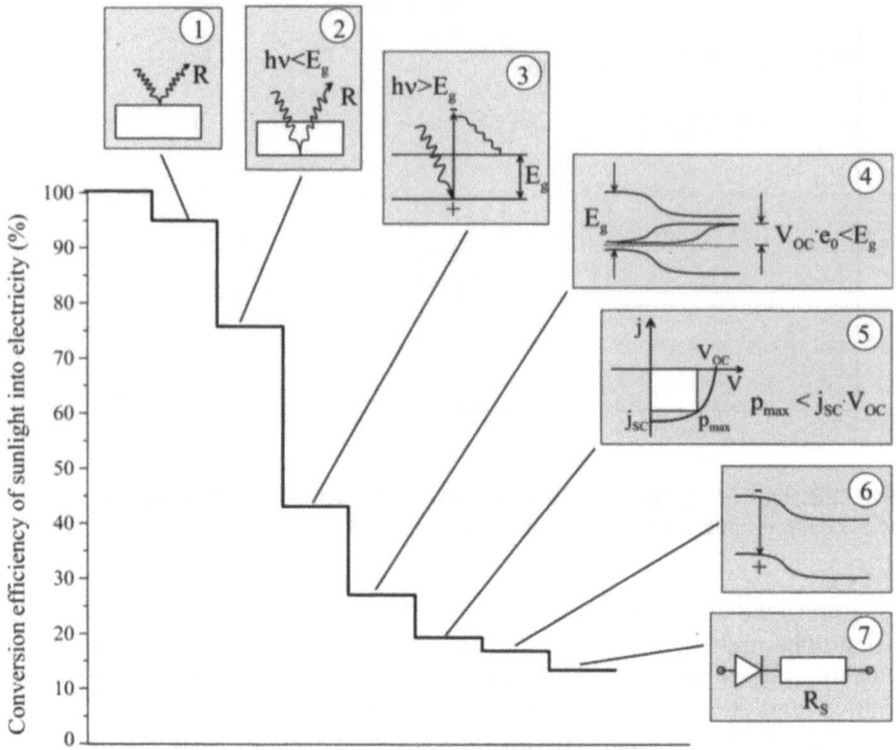

Fig. 2.4. The effect of individual loss processes on the efficiency of a typical solar cell. The various mechanisms are explained in the text

wafer solar cells. The thin-film solar cells made of a-Si, CdTe or CIS all rely on layers of transparent conductive oxides (TCOs) as the front contact.

An important parameter of a solar cell is the conversion efficiency, which is defined as the ratio of the electrical power producedto the incident solar power. The latter is usually at best $1\,\mathrm{kW/m^2}$. Figure 2.4 illustrates the many physical and technological loss mechanisms that result in the typical 14 % conversion efficiency of present-day industrial crystalline silicon solar cells shown here as a representative example. As we have seen earlier only photons of sufficient energy can be utilized in the conversion process. Others are not absorbed and are therefore transmitted (process 2). High-energy photons also produce only one electron-hole pair per photon; the surplus energy is converted into heat (process 3, see also Fig. 2.1). These two physical loss mechanisms alone reduce the conversion efficiency below the 50 % level. In addition there is imperfect coupling of useful light into the cell due to reflections from the front metal grid and the remaining front surface (process 1). The open-circuit voltage that can be obtained is physically limited to values less than the bandgap voltage (process 4). In our analysis we are now already down to the 25 % efficiency level, the highest conversion efficiency that has ever been reported in a small-area laboratory solar cell. Another inevitable loss mechanism is shown in process 5. Only 80 % of the product $I_{sc} \times V_{oc}$ can be extracted at the maximum power point P_{mpp} since the I-V curve is never perfectly rectangular for physical reasons (Fig. 2.3). Recombination losses due to photogenerated carriers not reaching the electrical contacts are illustrated in process 6. The electrical series resistance R_s (process 7) in the cell itself, its contacts, and in the external circuitry lead, if not properly designed, to another loss in power delivered to an external load. The latter two mechanisms are strongly technology-related and depend on the quality of the starting material and the processing.

2.1.1 The Standard Screen Printed Crystalline Silicon Solar Cell

Standard crystalline silicon wafer dimensions in production are 5 in, square ($12.5 \times 12.5\,\mathrm{cm}^2$) or quasi-square, with a starting thickness of around 300 µm. The trend is toward larger ($15 \times 15\,\mathrm{cm}^2$) and thinner (250 µm) wafers. The wafers sawn from a single crystalline ingot or multicrystalline block are usually wet-chemically etched to remove sawing damage. In this process about 20 µm are removed on each side. This is followed by a cleaning procedure before a shallow heavily doped emitter is produced by driving phosphorus into the boron-doped silicon wafers (see Fig. 2.2) in a belt furnace at high temperature (900 °C).

In monocrystalline silicon wafers with $\langle 100 \rangle$ crystal orientation, the surface can be texture-etched in an anisotropic wet-chemical etching solution (i.e. where the etch rate depends on crystal orientation) prior to phosphorus diffusion to produce a random array of small pyramids (height 1–10 µm) which effectively reduces the fraction of reflected light (see Fig. 2.5). The surface can then be passivated by growing a layer of silicon oxide or nitride on it. The silicon nitride, when grown from nitrogen- and hydrogen-bearing gases at low temperature in a glow discharge, leads to hydrogen incorporation in the silicon, which effectively passivates defects both at the surface and in the bulk of the solar cell.

Electrical junctions are usually added to the solar cell by printing pastes that contain metal particles through a screen onto the surfaces of the wafer [Van 1986]. This process creates the required metallization pattern on both sides of the cell (Fig. 2.5). In addition to 5–10 µm silver or aluminum particles, the metallization pastes also contain low melting point glass powder that will etch through the nitride and into the silicon surface during the so-called firing step, and thus provide adhesion between the metal particles and the wafer. Organic binders keep the active powder in suspension and solvents are added to adjust the viscosity during the printing step. Those two components are removed when the printed cell is initially dried and then fired at high temperatures.

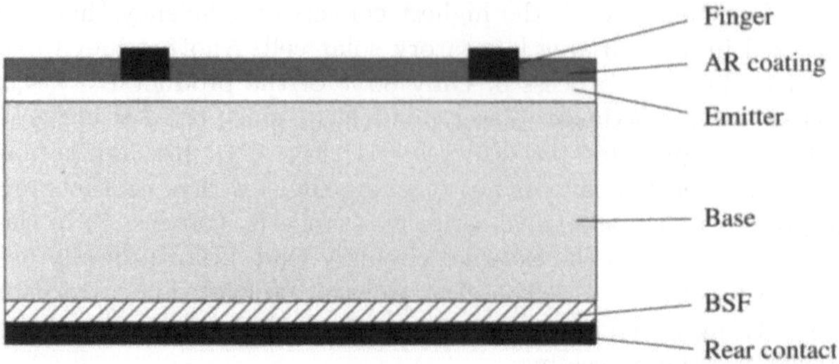

Fig. 2.5. Schematic view of a typical (screen-printed) standard crystalline silicon solar cell. Finger = front side electrical contact, BSF = back side electrical field (highly doped region which reduces energy losses of excited electrons at the contacts). See also Fig. 2.2

Besides the advantage of high metal usage, the screen printing method also has several disadvantages. For technical reasons the fine-line front metal grid finger width is limited to about 100 µm, which determines the minimum optical reflection losses achievable. The particle nature of the metal contained in the paste leads to incomplete metal coverage, with comparatively low finger conductivities and high contact resistivities. In order to minimize the latter effect a high phosphorus surface concentration is required, whitch on the other hand limits the photocurrent generated by such a heavily doped emitter. In addition, screen print pastes are comparatively expensive. Despite these drawbacks this method has developed into the standard metallization technique widely used in industry. This technique works well on sawn wafers with a well-defined surface. For EFG ribbon wafers (see below), which are drawn directly from the melt and are therefore not completely flat, an adapted metallization procedure had to be developed.

Frequently a heavily aluminum- or boron-doped p-type layer is created at the back side of the wafer. Through this process the surface recombination of electrons reaching the back surface by diffusion is strongly reduced. Especially in thin wafers (of good quality) this is a simple and attractive way to enhance the efficiency of the device.

As the final cell manufacturing step, the solar cell parameters of each individual cell are measured and the cells are classified according to their performance.

The standard screen printing process described above results in 13–15 % efficient polycrystalline and 15–16 % efficient monocrystalline silicon solar cells. The respective module efficiencies which include the total module area, are typically 2 % below the cell values, mostly because of packing density effects. The efficiency gap between mono- and polycrystalline silicon is rapidly closing at present, due to improvements in both material quality and production technologies for the latter technology.

Despite decades of industrial production experience with this technology, fully automated continuous in-line manufacturing facilities have so far not been realized. Key issues in processing are wafer handling, which is of increasing importance due to cost-driven decreasing wafer thicknesses, and the implementation of cost-effective efficiency-enhancing technologies in the production environment.

2.1.2 High-Efficiency Silicon Solar Cells

In order to significantly increase the efficiency of a crystalline silicon solar cell, a more weakly doped emitter has to be implemented with a resulting higher resistivity of the emitter layer. This requires more narrowly spaced front metal fingers and – to keep the shadowing losses small – finger widths smaller than can be realized by the screen printing technique. An elegant alternative concept has been developed in the so-called buried contact solar cell (Fig. 2.6). Here 30 µm-wide grooves or trenches are produced in the surface of the wafer by laser cutting. If this cutting is performed through a masking oxide or nitride layer, the resulting local opening can be used to advantage to apply a second heavy n^{++} diffusion with the rest of the surface being masked from the diffusion. This so-called selective emitter combines the relatively good light conversion properties of the comparatively weakly doped n^{+} emitter between the contacts with the good contacting properties of the heavily doped n^{++} emitter at the contact.

The laser-cut grooves are filled with metal in an electroless plating process. The standard process deposits nickel from a nickel salt solution autocatalytically on exposed semiconductor surfaces but not on the surfaces of the dielectric. Therefore nickel is only deposited in the grooves and not on

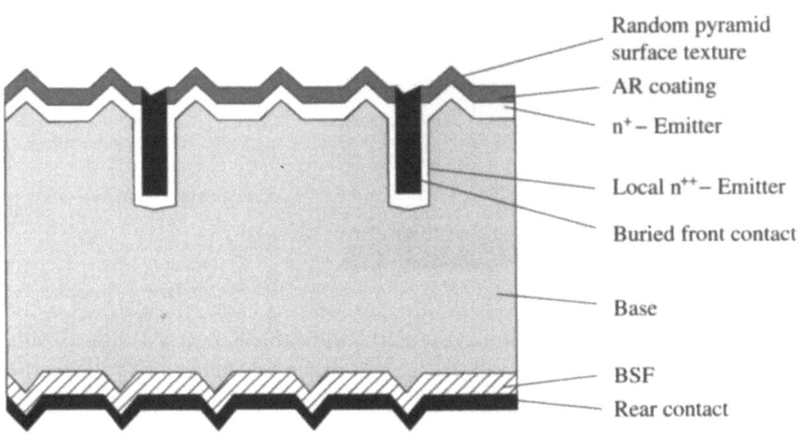

Fig. 2.6. Buried contact crystalline silicon solar cell. See also Figs. 2.2 and 2.5

the other – oxide or nitride covered – surfaces. These buried contact cells are in industrial production where cell efficiencies of up to 17% are reached in CZ-silicon (see below).

The "passivated emitter rear locally diffused" PERL cell concept shown in Fig. 2.7 has some additional features realized by microelectronics processing techniques such as vacuum evaporation, and most notably photolithographic pattern generation methods.

The first feature is the 10 μm-deep so-called inverted pyramids that are generated on the single crystalline surface of a ⟨100⟩-oriented wafer by producing a finely patterned masking oxide layer and a subsequent selective etching of the exposed silicon surface in an anisotropic wet-chemical etching solution. Such a textured surface provides excellent antireflection and light trapping properties.

The next feature is the 10 μm-wide front metal fingers that are produced in a so-called lift-off process where the whole front surface is covered with metal evaporated onto a patterned layer of photoresist. This is subsequently removed, thereby taking most of the metal with it and leaving only the fine strips of finger metal in precisely aligned positions. These fingers are usually thickened in a galvanic plating process.

The third feature is that both front and rear surfaces are almost completely (with the exception of tiny openings made again by photolithographic patterning) covered by a high quality high temperature so-called thermal oxide slowly grown by exposing the silicon wafer to a dry oxygen ambient atmosphere. A thick layer of this thermal oxide provides excellent surface passivation.

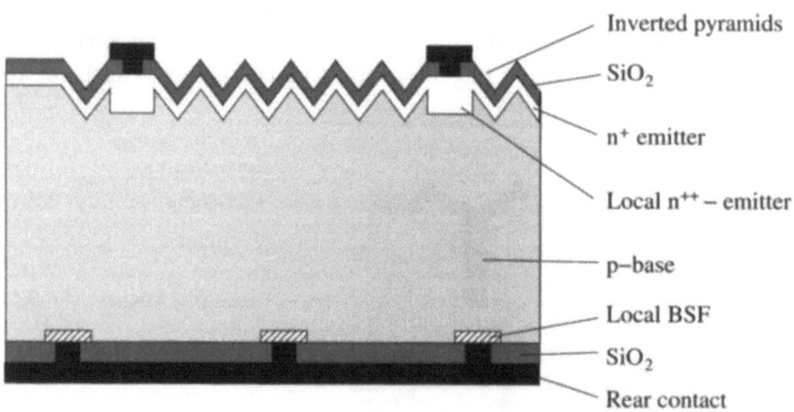

Fig. 2.7. Passivated rear locally diffused (PERL) high efficiency crystalline Si solar cell. SiO_2 is used for the elctrical passivation (in order to reduce energy losses) of the surfaces. See also Fig. 2.6

The fourth feature is the photolithographically defined generation of small islands of boron-diffused back-surface field in places just where the backside metal penetrates the otherwise oxide covered surface.

The fifth and last new feature is the back side mirror with near perfect reflection properties that is realized by the combination of thermal oxide and an evaporated rearcontact Al layer.

The incorporation of all these features has resulted in record cell efficiencies above 24 % on small-area solar cells prepared on selected prime quality specially purified so-called float-zoned (FZ) wafers. Present research aims at developing low-cost high throughput techniques that replace the photolithographic masking steps, but keeps up the cell efficiency in the 20 % range. A first breakthrough has been obtained by demonstrating 20 % solar cell efficiency industrial material (CZ, $10 \times 10\,\text{cm}^2$) obtained with only two masking steps [Glu 2000], which is a significant reduction from the previous six masking steps.

2.1.3 Module Manufacturing

The sorted and classified crystalline silicon solar cells require encapsulation for protection purposes. This encapsulation provides mechanical rigidity to support the brittle cells and their flexible interconnections. It also provides protection from mechanical damage caused by hail and it protects the metallic contacts and interconnections from corrosive elements in the atmosphere. Finally it provides electrical isolation of the voltages generated by the panel which may be up to a 100 V (see below). It is not the cells themselves but the durability of the encapsulation that determines the ultimate operating lifetime of the module. Additional features an encapsulation scheme must possess are stability under ultraviolet (UV) light, tolerance to temperature extremes and thermal shocks without stressing cells to fracture, resistance to abrasion in certain climates, ability to keep the cell temperature low to minimize power loss and of course above all low cost.

First of all the individual cells have to be manufactured into strings which, depending on module size, are composed of up to 18 cells (typically 9 cells). These are not magic numbers, but rather have to do with the fact that since each cell delivers at best 0.6 V, 36 cells are usually connected in series in a module to deliver the voltage necessary to charge the typical 12 V battery system.

Cell interconnection normally occurs through soldering flat copper strips to the front of one cell and the back of the adjacent cell. This is no tight interconnect but some stress relief loop is incorporated in between the cells to compensate differences in thermal expansion in the finished module (Fig. 2.8). Several strings are now further interconnected to form a module-sized array of cells. Standard modules contain multiples of 36 cells (up to 216 cells) which are then generally connected in parallel to deliver more current rather than more voltage.

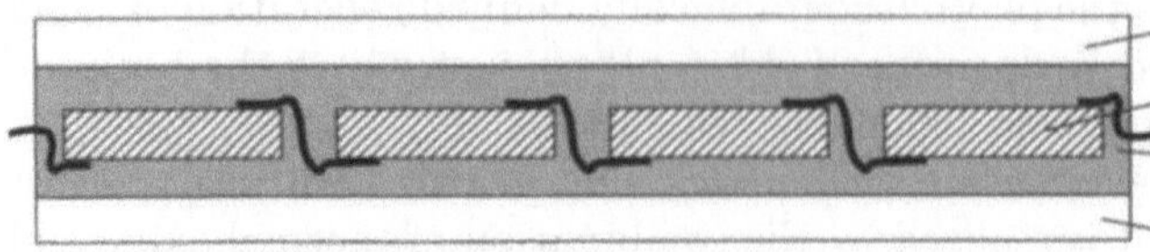

Fig. 2.8. Laminated crystalline silicon solar module. The electrical interconnections between individual cells are shown as black line segments

The module is then made by forming a sandwich of hail-resistant specially tempered (hardened) front cover glass, a thin sheet of ethylene vinyl acetate (EVA), the cell string array, another sheet of EVA, and the back encapsulant, which may be either another glass plate or the presently preferred choice, a sheet of TEDLAR (Fig. 2.8). The latter is the trademark of a polymer composite material that incorporates aluminum foil as a moisture barrier. The laminate is now ready for lamination, a process well known from car glass manufacturing, which involves the application of vacuum to remove air trapped between the layers and the subsequent simultaneous application of both pressure and heat to melt the EVA layers. When carried out properly this process results in a well-sealed final product with an expected and presently warranted lifetime of 25 years.

The module is completed by attaching a so-called junction box to it that carries connectors for external wiring as well as so-called bypass diodes which are safety devices. These bypass diodes effectively block power from the rest of the module being dissipated in a partially shadowed cell. This might result from a leaf falling on the module which in turn might lead to a so-called hot spot (i.e. local heating) with resulting delamination or even complete destruction of the cell in question. A frame may or may not be attached to the module for installation purposes.

2.1.4 Thin Film Solar Modules

A radically different approach from the separate handling of self-supporting crystalline silicon solar cells is taken in the thin-film solar cell technologies. Here very thin layers a few micrometers thick – a fraction of the thickness of a human hair – are deposits from the gas phase onto a suitable foreign substrate. In the most successful technologies these depositions take place at such low temperatures that window glass or in some cases metal foils can be used as the film-carrying material [Aul 1995]. However, these low film growth temperatures lead in the case of amorphous silicon to non-crystalline or in the case of cadmium telluride (CdTe) or copper indium diselenide (CIS) materials (see below) to small-grained polycrystalline films, where the size of the crystallites is typically of the order of the film thickness or smaller. These layers contain structural and electronic defects and are therefore usually of lower electronic quality than the 1 cm or 10 cm-large crystals involved in the crystalline silicon solar cell technologies.

The gas-phase deposition processes have been scaled up such that substrates of module size (up to a square meter) can be coated with thin films. If this large area were manufactured into a single solar cell, it would lead to impractically large currents. Therefore, the whole area must be divided into many individual cells that are about a centimeter wide. On an electrically insulating glass substrate a cell interconnection procedure can be comparatively easily integrated into the cell manufacturing process. In order to add the voltages of the individual cells, each rear contact of the previous cell must be connected to the front contact of the next adjacent solar cell. In general the process sequence to manufacture thin-film modules is quite similar for different semiconductor systems, although the materials used and the production methods employed vary quite considerably.

In the first step (Fig. 2.9) the glass substrate is completely coated with the electrode. This may be a non-transparent molybdenum layer or a transparent conductive oxide (TCO) like tin oxide (SnO_2) or zinc oxide (ZnO,) which is subsequently patterned, i.e. narrow lines of material are removed using laser or mechanical scribing. In the second step the solar cell base and emitter layers are deposited over the whole substrate and patterned using similar techniques as in step one. This patterning process must ensure, however, that the underlying first electrode is not removed.

This step is followed by the depositing and patterning of the second electrode, which may be a thin layer of a metal like silver, aluminum, or a metal oxide like ZnO. This last patterning step usually cuts through the second electrode, as well as the base and emitter semiconductor layers, leaving the first electrode intact (as illustrated in Fig. 2.9).

Finally the thin-film device is encapsulated using materials similar to those of crystalline silicon technology, like ethylene vinyl acetate (EVA) and a sheet of Tedlar to protect the circuit from the environment. The rough processing sequence sketched here serves well to show the similarities between the different thin-film technologies, indicating that the solar cell base and emitter layers, usually receiving major attention, are only part of several other important processes.

In fact, closer analysis of the process reveals that depending on the technology, roughly 15 process steps are necessary to complete a module. With a targeted overall production yield of over 90 % this requires individual process steps with yields above 99 %. Accomplishing this on large areas with fast homogeneous and high electronic quality deposition leading to module efficiencies high enough for low-cost applications has been extremely demanding, and is one important reason why thin-film technologies so far play no major role in photovoltaic power generation.

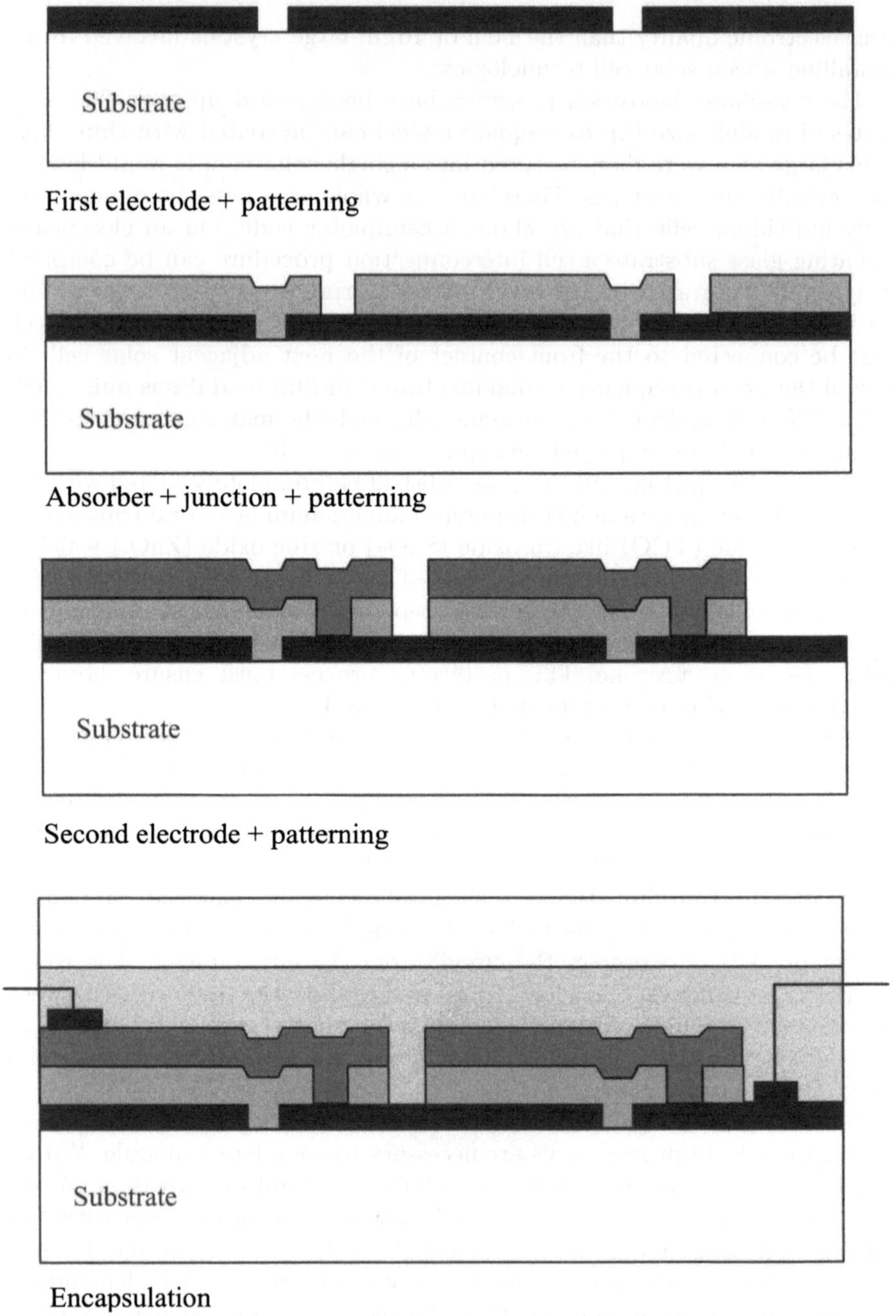

Fig. 2.9. Processing sequence for monolithically series-interconnected thin-film solar modules. The photovoltaic light energy conversion is done in the absorber/junction system. Patterning allows for the series interconnection of the individual cells

2.1.5 Amorphous Silicon (a-Si)

Among the thin-film materials for solar cells amorphous silicon (a-Si) is by far the most extensively researched material with the largest manufacturing experience [Guh 1999]. Solar cells from amorphous silicon are mainly used in low-power devices for indoor applications like calculators and watches. The rapid development of this technology has benefited strongly from the parallel development of other electronic devices like thin-film transistors in liquid crystal active matrix displays, photosensitive layers in photocopying machines, and other sensor applications. This synergistic boost is quite like the positive effect that silicon microelectronics has had and still has on the development of crystalline silicon solar cell technology. This is to a certain extend a weak point with CdTe and CIS solar cell technologies (see below), where no other widespread applications outside photovoltaics have yet been found.

The two major hurdles that must be overcome to manufacture low-cost a-Si solar cells are the limited efficiency and the photodegradation intrinsic to the material. These limiting factors are the main reaspon why the introduction of a-Si modules into photovoltaic power applications has not occurred so far.

Thin layers of amorphous silicon are usually deposited in a so-called glow discharge in a silicon-bearing gas – typically silane (SiH_4). These layers behave like a supercooled liquid and are not thermodynamically stable. Given sufficient temperature and time (e.g. 500 °C for several hours), these layers will crystallize spontaneously into very fine-grained polycrystalline silicon.

In order to obtain electronically viable material that can be doped and formed into working devices, hydrogen must be incorporated the irregular amorphous structure. This incorporation leads to a significant reduction of internal defects by a passivation effect. The defects would otherwise prevent any useful electronic application.

The advantage of the amorphous structure is that it absorbs light much more strongly than its crystalline counterpart and is therefore suitable for thin-film solar cell formation. However, the disadvantage is that its electronic properties such as carrier mobilities and photogenerated carrier lifetimes are much inferior to its crystalline brother.

This requires the formation of a so-called pin-type solar cell structure (Fig. 2.10) where most of the light is absorbed in the so-called intrinsic (i) – or undoped, highly resistive – region sandwiched between extremely thin p- and n-type layers. This structure has a junction and therefore an electric field region extending throughout most of the solar cell, and this electric field helps to separate and collect the charge carriers which would otherwise recombine rather effectively. The electric field also helps to reduce the degradation mechanism which is based on a recombination-induced defect creation, which in turn weakens the electric field in the i-layer. In order to

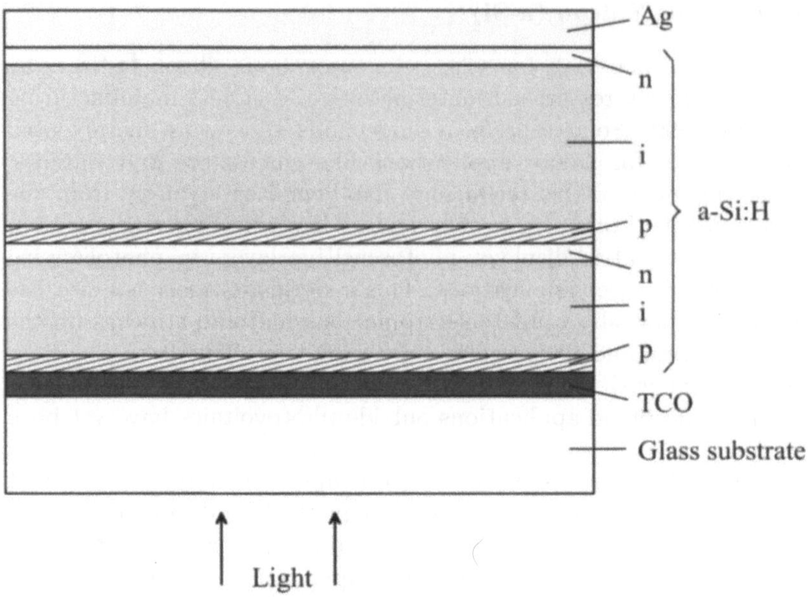

Fig. 2.10. Schematic cross section of an a-Si tandem thin-film solar cell. Ag = elctrical silver contact; n, i, p = different amorphous silicon materials, TCO = transparent electricity conducting oxide (front contact)

have a strong electric field, a thin i-type absorber is required, which leads to a loss in absorption and a reduction in solar cell efficiency.

This problem is eliminated – at the cost of much more complex technology – by depositing two or three solar cells on top of each other. These tandem or triple-junction devices require heavily n- and p-doped layers that form a so-called tunnel junction between solar cells as well as current matching between the individual cells in a stack for optimized operation. The latter leads to the fact that the top cell (the one exposed directly to the sun) has to be made somewhat thinner than the bottom cell for the latter to receive enough light to produce the same current as the top cell. Another possibility is to reduce the bandgap of the bottom cell by alloying the Si with Ge, which leads to better utilization of the photons in the solar spectrum.

The best small-area laboratory triple-junction devices have an efficiency of up to 14 %. Large size single junction a-Si industrial modules are currently available with stable 4–5 % efficiency and tandem modules with stable 6–7 % efficiency.

2.1.6 Copper Indium Diselenide (CIS)

Polycrystalline $CuInSe_2$ (CIS) and related chalcopyrite compounds are interesting candidates for photovoltaic power applications [Schock 2000]. When

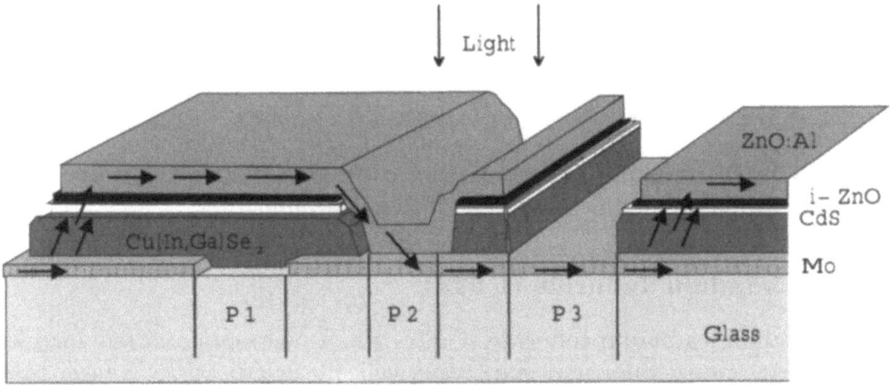

Fig. 2.11. CuInSe2 (CIS) thin-film solar cell structure. The arrows indicate the electron current flow through the two interconnected cells shown in the figure. ZnO:Al = transparent electrode, CdS = cadmiumsulfide layer, Mo = metallic molybdenum rear contact

replacing some of the indium in CIS by alloying it with gallium, the resulting Cu (In, Ga) Se_2 (CIGS) has an optimum bandgap with solar cell efficiencies around 18 % demonstrated on small-area laboratory devices.

All high-efficiency CIS devices (Fig. 2.11) use a substrate of conventional (soda-lime) float glass coated with a 1 μm-thick layer of sputtered molybdenum, a 2–3 μm-thick CIGS absorber, a cadmium-sulfide (CdS) layer some tens of nm thick, usually grown in a chemical bath deposition, and a 1–2 μm-thick transparent sputtered zinc oxide (ZnO) electrode. Two approaches for CIGS have been successfully employed in the production of high-efficiency devices – co-evaporation of the elements and the sequential sputtering of copper and indium layers followed by selenization in an $H_2$2Se ambient.

The sodium in conventional window glass appears to have a very beneficial effect on the crystal quality and the electronic properties of the absorber layer into which it diffuses through the molybdenum electrode during high-temperature processing. The diffusion of sodium into CIS strongly depends on the chemical composition of the glass surface, the morphology and thickness of the molybdenum film, and the deposition conditions of the CIS absorber. These parameters must be controlled precisely over the large module area in order to develop high-yield, low-cost manufacturing processes. One promising approach is to use suitable barrier coatings on the float glass to obtain a well-defined surface composition, and to deposit a sodium compound onto the molybdenum electrode to ensure controlled sodium diffusion.

The highest solar cell efficiencies are apparently related to the formation of a thin layer of n-type $CuIn_3Se_5$ at the surface of the p-type $CuInSe_2$ absorber layer, and this requires a slightly indium-rich composition during growth. Nevertheless an additional extremely thin layer of n-type cadmium sulfide deposited in a chemical bath dip is also required. This buffer layer provides some protection against interaction with the oxygen during ZnO deposition, as well as beneficial surface passivation.

In comparison with a-Si and CdTe (see below), the above-described manufacturing processes of CIS may be considered to be somewhat more complex, and therefore throughput and production yield are key issues. The first commercially available modules have an efficiency around 10 %.

2.1.7 Cadmium Telluride (CdTe)

Polycrystalline cadmium telluride (CdTe) has a near-optimum bandgap for solar energy conversion, and small-area cell efficiencies of 16 % have been demonstrated on laboratory-type cells [Mey 2000]. A whole range of deposition techniques has been successfully used [Bon 1992], which is due to the fact that above 400 °C the stochiometric CdTe compound is the only stable phase; its slight Cd deficiency leads to a native p-type doping. CdTe solar cells are prepared on TCO coated glass substrates onto which a 100 nm thin n-type CdS layer has been deposited (Fig. 2.12). An attractive CdTe deposition technique is close space sublimation, where CdTe vapor is produced in a reduced pressure ambient from a solid source at 700 °C; this vapor condenses across a small gap on the solar cell substrate kept at about 500 °C. This technique allows high growth rates of 5–10 μm/min with a typical absorber thickness of 3 μm and a high material utilization of about 70 %.

Among the critical issues in CdTe technology is the need for post-deposition treatment using $CdCl_2$, which seems to be essential to obtain high efficiency. This treatment leads to grain size enhancement and passivation of grain boundary defects. Furthermore, obtaining a stable low cost and low-resistance rear contact under high-yield manufacturing conditions is a challenge.

Large-area module efficiencies in the 8–10 % range have already been achieved in a pilot production environment.

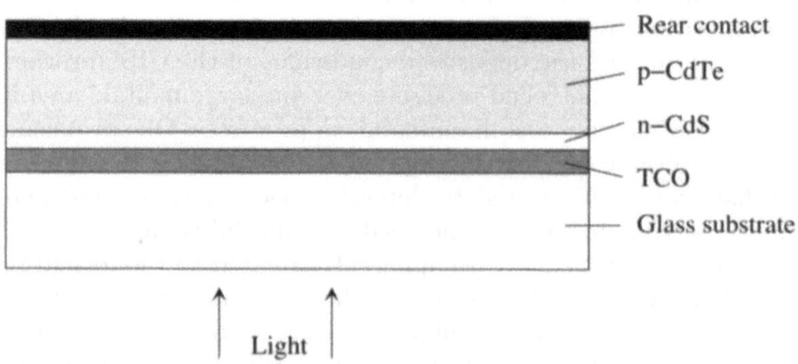

Fig. 2.12. CdTe thin-film solar cell structure. See also Fig. 2.10

2.1.8 Raw Materials and Resources

Silicon is after oxygen the second most abundant element in the earth's crust, where it does not occur in its elemental form but only in the form of compounds, mainly silicon dioxide and silicates. It therefore has to be reduced to elemental silicon and this is carried out in electric arc furnaces at 2000 °C using carbon as a reducing agent. This process – which of course releases CO_2 – uses about 10 kWh of electricity per kg of technical (metallurgical grade: mg) silicon produced, which is 98–99 % pure. Of this material 700 000 t per year are produced, mainly in Norway and Canada, where cheap and CO_2-neutral hydroelectric power is available. The major fraction of this material goes into metallurgical applications (e.g. processing of aluminum) as well as into the silicone industry. Only a small percentage of it (20 000 t/yr) is purified for use in the microelectronics industry. For this purpose the mg-Si is mechanically crushed into silicon dust which is then reacted with anhydrous HCl to form chlorosilane compounds, which are easily purified by fractionated distillation. The highly purified trichlorosilane $SiHCl_3$ is then pyrolytically decomposed to solid Si using H_2. This so-called electronic grade silicon (eg-Si) is further processed in a melting step to single crystalline Czochralski (CZ) ingots, which are sawn into wafers and used in the microelectronics industry. The out off spec waste material (which is at present about 2 000 t/yr) from this process is the main source of raw material for the photovoltaics industry. This fact may produce a problem. If the market for crystalline silicon solar cells continues growing at the present rate, a serious silicon feedstock shortage will be felt at the latest by the year 2005. Until then an independent sustainable supply of low-cost so-called solar grade silicon (sog-Si) must be developed.

An increasing fraction of solar cells are produced from multicrystalline silicon that has been block-cast by directionally solidifying molten silicon in a container. These typically $50 \times 50 \times 30\,cm^3$ large blocks are sawn into smaller columns, which in turn are wire sawn into wafers. In this process about half of the precious silicon is transformed into sawdust.

An interesting alternative wafer producing process that has reached the industrial stage is the so-called edge-defined film-fed growth (EFG) technique. Here liquid Si is drawn by capillary forces through an arrangement of graphite dies. Such a process results in the growth, for example of octagonal hollow tubes (4.5 m long, 26 cm diameter) which can be processed directly into wafers by laser cutting. This process results in much greater silicon material utilization than the standard wafering technique.

The silane used in the formation of amorphous Si modules as well as the $SiHCl_3$ used to produce crystalline silicon thin-film solar cells (see below) are generated in large quantities in the silicon purification scheme described above. The material supply for these types of solar cells therefore does not run into any shortage problems.

The most important semiconductor materials used in the preparation of CIGS solar cells are Cu, In, Ga, Se, and to a much lesser degree, CdS. The least abundant and therefore most critical elements are In and Ga. For example the present global annual production of In is 134 t. If this could be completely used in CIGS solar cell production the amount would be sufficient to generate 4.5 GW/yr. This figure represents the total market size that could be reached in about 20 years if the market continues growing at the present rate (see Sect. 4.3.3).

Cadmium is produced as a byproduct in the mining of Zn and is used mainly in the fabrication of NiCd rechargeable batteries, which in the US alone uses 4 400 t/yr of Cd. If this amount were used to generate CdTe solar cells it would suffice to supply the expected total market for the next 30 years. Tellurium is much more rarer; it is only produced at 350 t/yr which would supply at most the complete world PV market for the next 15 years if the mining rate remains unchanged (see Sect. 4.3.3).

2.2 Emerging Solar Cell Technologies

In the following we will mention three other emerging solar cell technologies that may play a significant future role in the photovoltaic power market.

2.2.1 Solar Cells Based on GaAs and Other III-V Compounds

Gallium arsenide (GaAs) is, after crystalline silicon, the second most important semiconductor material. It is employed in high frequency and optoelectronic devices such as light-emitting diodes (LEDs) and lasers. Also GaAs has an optimum bandgap for solar energy conversion [Bet 1999], which has resulted in solar cell efficiencies above 25 %, the highest value ever achieved in a single junction structure under non-concentrating sunlight conditions.

So far the high price of GaAs wafers, which are a factor 50 more expensive than Cz-Si wafers of comparable size, has prohibited its use in conventional terrestrial PV applications. However, the development of single crystalline thin-film GaAs solar cells on somewhat (a factor 5 with respect to GaAs) cheaper Ge substrates has led to the fact that they currently successfully compete with and replace crystalline silicon solar cells in space applications such as powering telecommunication satellites. The advantages of GaAs solar cells are their superior radiation hardness (i.e. stability against extraterrestrial bombardment with solar protons and electrons), which lead to a longer lifetime and a higher solar cell efficiency. This leads to smaller systems with a higher power-per-weight ratio. In this market, which is about 1 % of the current terrestrial photovoltaic market, GaAs has conquered a share of about 50 %.

A major advantage of this class of materials is that by combination with other elements in groups III and V of the periodic table such as Al, In, N, P, and Sb, a whole range of semiconductors with smaller and larger bandgaps

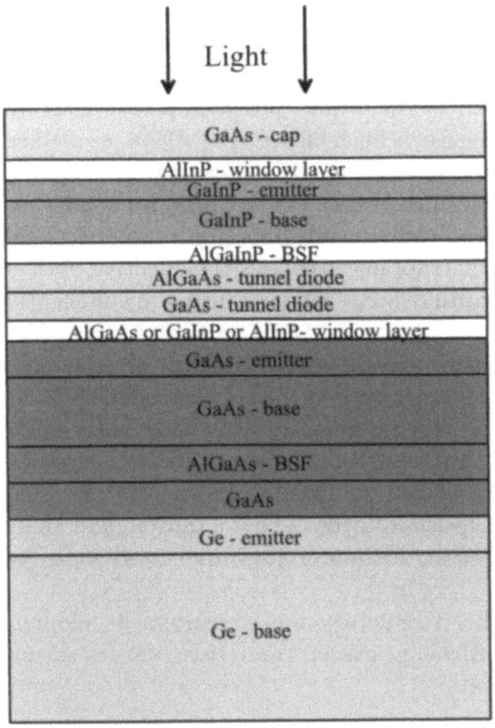

Fig. 2.13. Multiple junction III/V solar cell. The upper two layers constitute the transparent front contact. A tunnel diode is used to interconnect the upper and the lower photovoltaic cell of the tandem structure. Without the germanuim basis the whole photovoltaicly active structure has a thickness of some μm

can be generated, which can be combined in multiple-junction solar generators (Fig. 2.13) that much more effectively make use of the energy contained in the solar spectrum. As an example, conversion efficiencies above 27 % have already been achieved in GaInP/GaInAs tandem devices under standard terrestrial irradiation conditions.

For terrestrial applications the development of an optical concentrator technology is an attractive option with high cost-reduction potential [Swan 2000]. Here direct sunlight is focused onto a small solar cell by an optical system composed of lenses or mirrors. In a system that concentrates sunlight by a factor 100–1 000, the cost of the solar cell is expected to become a minor part of the total system cost. An attractive feature is that the solar cell conversion efficiency increases with light intensity (i.e. with concentration) by something like 10–20 % relative to the above-mentioned concentrations, provided the cell temperature can be kept constant. In such systems cell efficiencies above 30 % have already been demonstrated under concentrations up to 1 000 (which represents an irradiation of 10 W per cell of typical size 0.1 cm^2).

2.2.2 Crystalline Silicon Thin-Film Solar Cells

Crystalline silicon solar cells (Fig. 2.14) have the potential for high efficiency even for thicknesses in the range well below 50 μm, provided a crystal grain size much larger than the cell thickness can be achieved [Ber 1999], as well as an efficient light-trapping scheme incorporated in the cell design, for example by some surface texturing. Light trapping is essential in thin-film solar cells because in comparison to a-Si, CIS, and CdTe, crystalline silicon is a relatively poor absorber of solar photons. Light trapping increases the effective optical path length of photons in the photovoltaic device, thus resulting in sufficiently high photon absorption.

One approach in this field is to grow a 10–30 μm silicon layer directly on a suitable substrate from a silicon-bearing gas such as the trichlorosilane $SiHCl_3$ used in the purification of crystalline silicon. In order to achieve sufficiently large growth rates of about 10 μm/min, the substrate has to be heated to temperatures above 1 000 °C. Sufficiently large grain sizes may be obtained by melting and recrystallizing the deposited layer. These requirements alone severely limit the range of useful substrate materials, which also must be produced at low cost.

Nevertheless the potential of this technology has already been demonstrated by achieving a solar cell efficiency grater than 19 % with a 45 μm layer grown on an insulating substrate.

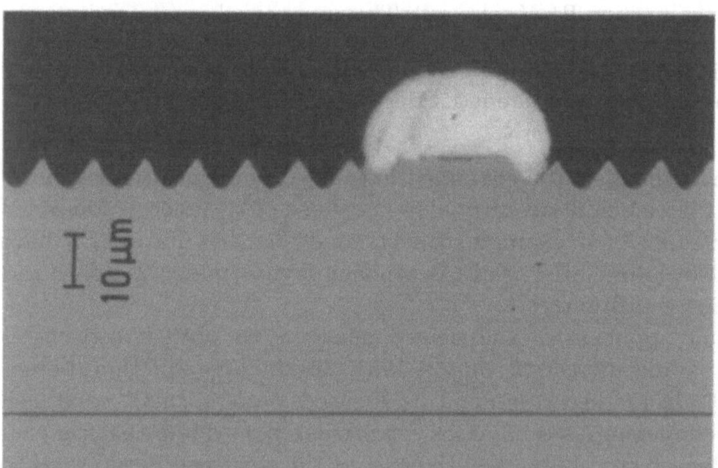

Fig. 2.14. Cross section through a thin film crystalline silicon solar cell. The textured surface and one finger of the front contact can be seen in the upper part of the picture. The horizontal line in the lower part indicates the interface betwen the substrate and the active solar cell

2.2.3 Organic and Nanocrystalline Solar Cells

The conversion of photon energy into electronic energy can also be realized by means of organic dye molecules. In this conversion scheme the absorption of a photon results in the generation of an energetically excited electron in the dye molecule. This charge carrier is then transferred to an electron transport medium in contact with the external negative electrode. This process leaves the dye in a positively charged (ionized) state. Via a similar but separate transport chain, an electron is then supplied from the external positive electrode to the ionized dye, thus bringing it again to an electrically neutral state. The recombination rate of excited electrons in such dyes is extremely high. Thus the dye molecules must be in direct contact with the electron-accepting transport medium in order to transfer the excited electrons in a high-speed process to a place where the recombination probability is considerably lower. This requires very large internal surfaces – covered e.g. with monolayer optical dye absorbers – in order to have sufficient light absorption in the cells.

In different solar cell schemes the absorbers are single molecules or polymers. Transport media can be realized as nanocrystalline semiconductor structures (cheap titanium dioxide particles sintered together), fullerenes, conducting polymers, etc.

Cells based on dye-sensitized nanocrystalline titanium dioxide structures (Fig. 2.15) have reached on small areas 11 % efficiency [Gra 2000]. Plastic (polymer) solar cells (Fig. 2.16) show efficiencies on small areas of the order

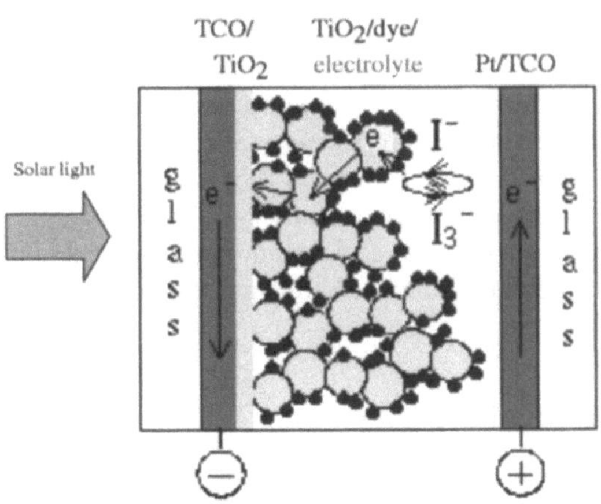

Fig. 2.15. Nanocrystalline dye-sensitized solar cell. Titaniumdioxide spheres (diameter typically 10 nm) are coated by dye molecules (black dots). Light absorption occurs in these molecules. Energetically excited electrons are subsequently injected from the dyes into the TiO_2 network. They then percolate in this medium to the elctrical front contact. The ionized dye molecules are discharged via an electrolytic electron transport (I-/I3-system). Do to the large surface of the TiO_2 matrix monolayers of dye molecules can be used in this photovoltaic conversion scheme

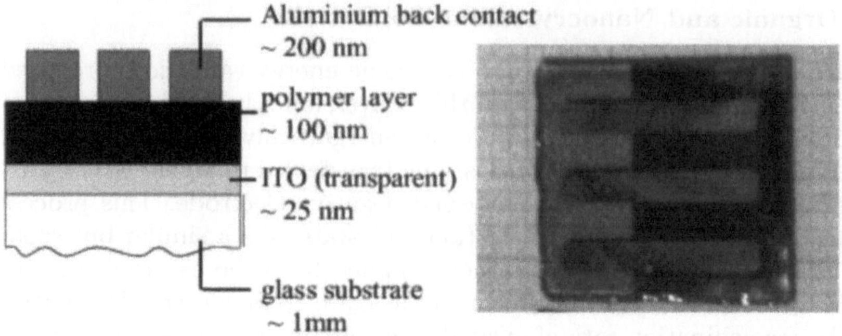

Fig. 2.16. Plastic solar cell. ITO = indiumtinoxide transparent electrode

of several percent. Efficiencies and the long-term stability of such cells have still to be improved considerably. On the other hand, these types of solar cells have, in principle, the potential for important cost reductions.

2.3 Photovoltaic Systems – Solar Electricity Supply

The modular nature of photovoltaic generators consisting of photovoltaic modules means that energy supply systems can be constructed over an extremely wide power range. The power spectrum ranges from a few mW for watches or calculators to kW systems as remote power supplies for mountaineering lodges or water pumps, and large central photovoltaic power stations in the MW range.

Photovoltaics are generally considered to be an expensive method of supplying electricity. On the other hand there is a growing market for this kind of energy technology all over the world. This indicates that solar electricity has entered areas where its application is already economically viable. In parallel with cost reductions, these areas will grow and extend to other fields of electricity supply. The current market breakdown of the various categories of applications is given in Fig. 2.17.

2.3.1 Different Types of Photovoltaic Systems

Photovoltaic systems are solar energy supply systems, which either supply power directly to electrical equipment or feed energy into the public electricity grid (Fig. 2.18).

Solar electric power supplies exist for appliances and small loads in the power range from several mW to several hundred watts. These power supplies are either grid-independent or remote from the grid. In these cases the reason for applying photovoltaics is generally their cost effectiveness. Other advantages besides the production of electricity are reliability, portability,

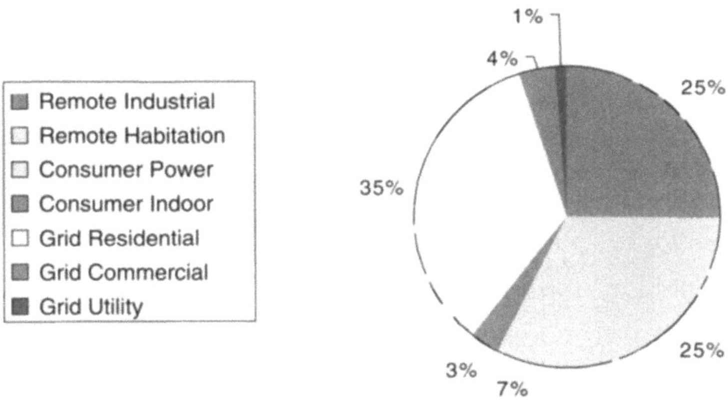

Fig. 2.17. Shipment of PV modules in 1999 [Strat 2000]

environmentally benign production of energy, etc. In many cases these advantages exist even if the consumer is situated near the grid. As can be seen from Fig. 2.17, almost 40 % of all modules shipped in 1999 are connected to the public electricity grid. These applications include both smaller, decentralized, grid-connected systems, with the solar modules typically installed on house roofs, and large, central, grid-connected systems, with large areas of solar modules usually mounted on dedicated support structures.

2.3.2 Off-Grid Solar Electric Systems

In very small systems such as watches or calculators, the solar generator consists of one or only a few solar cells. When more power is needed, the individual solar cells are connected to form solar modules. An energy storage unit is needed to bridge times when not enough light is available to power the appliance directly. Nickel-cadmium rechargeable batteries are used in most photovoltaically powered appliances, but lead-acid batteries and capacitors (so-called supercaps) can also be used. For caravans, boats, and week-end accommodation, special models of car batteries with particularly thick lead plates and special alloys for mobile application can be used. In photovoltaic systems for continuously occupied houses and daily charge/discharge cycles, the so-called "OPzS batteries" are most commonly used. They are characterized by very low self-discharge, and extremely good tolerance to cycling, and thus a long operating lifetime. For small systems, maintenance-free batteries with the electrolyte integrated in matting or in gel form are used. They cannot leak and can thus be operated in any position.

As lifetime costs for a standalone photovoltaic system are highly dependent on battery costs, careful selection of the best-suited technology for the respective application, operation, and battery management schemes are crucial to satisfactory use.

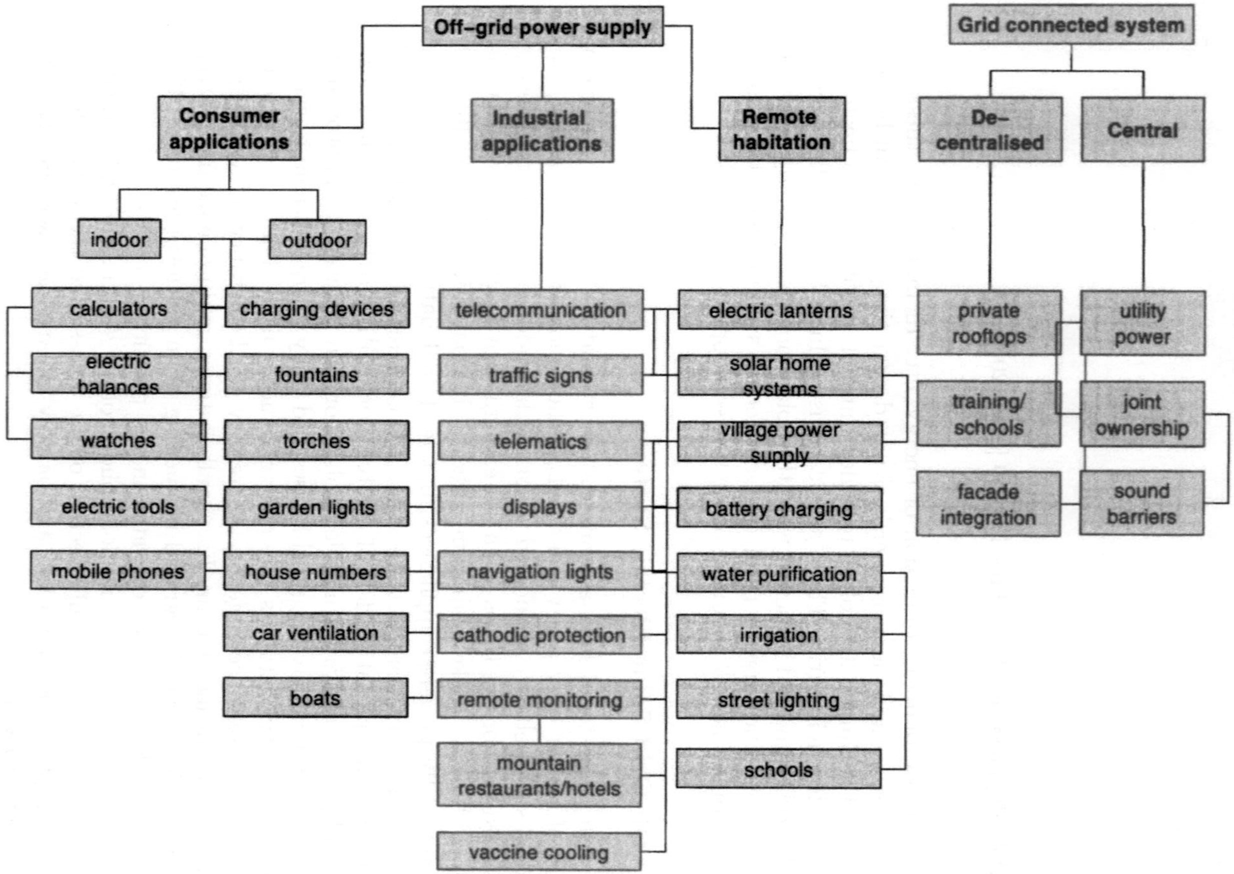

Fig. 2.18. Application areas of photovoltaic systems. The applications are subdivided into “off-grid” and “grid-connected” systems. While photovoltaics have manifold possibilities for being used in off-grid applications, the market for grid connected systems has been growing steadily in the last few years

A charge controller is therefore included between the solar generator and the battery to prevent it from being overcharged or deep discharged. The charge controller usually has a blocking diode, which prevents the battery from discharging during the night via the solar generator. A good charge controller has very low internal power consumption and includes a load cutoff switch which protects the battery against discharge.

Power conditioning may be needed to adapt the voltage level of the photovoltaic system to that of the load. For photovoltaically powered appliances, this is usually a DC/DC converter, which transforms one DC voltage to another. If AC voltage is needed by a consumer, an inverter must be used. This converts the DC voltage delivered by the solar generator or the battery to an AC voltage.

Consumer Applications Power systems for appliances and small devices where the energy supply and demand occur simultaneously do not need an energy storage unit (Fig. 2.19). Pocket calculators, fans, and electric letter balances are examples of such applications, [Amann 1996, Roth 1999-1, Roth 1999-2]. Systems can also be constructed without electricity storage if storable products are produced by the electricity consumer. Water supply systems are an important example of this type of photovoltaic application.

As one example of an appliance in the lowl power range, Fig. 2.20 shows an electric balance.

When an electric display is used, these devices normally are powered by dry-cell batteries. As energy demand and photovoltaic energy supply are very well correlated in this application (reading the LCD display is not possible without light), the system may be designed without any energy storage.

Besides environmental aspects, one fact is often neglected: the costs per kWh supplied with dry cells are extremely high. For example, a 1 000 mAh dry cell with nominal voltage of 1.5 Volt contains 1.5 Wh. Its costs of about 1 EURO per unit mean that the kWh stored in this way costs about 700 EURO/kWh. Not relying on a storage battery means system reliability is higher, negative environmental effects are reduced, and the comfort in using these systems is increased.

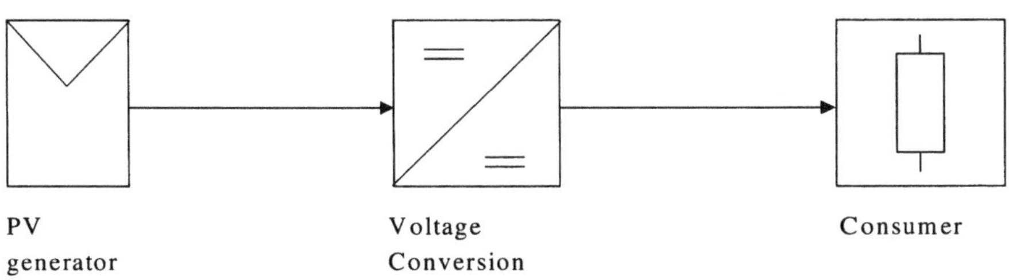

Fig. 2.19. Block diagram of a photovoltaic system without energy storage

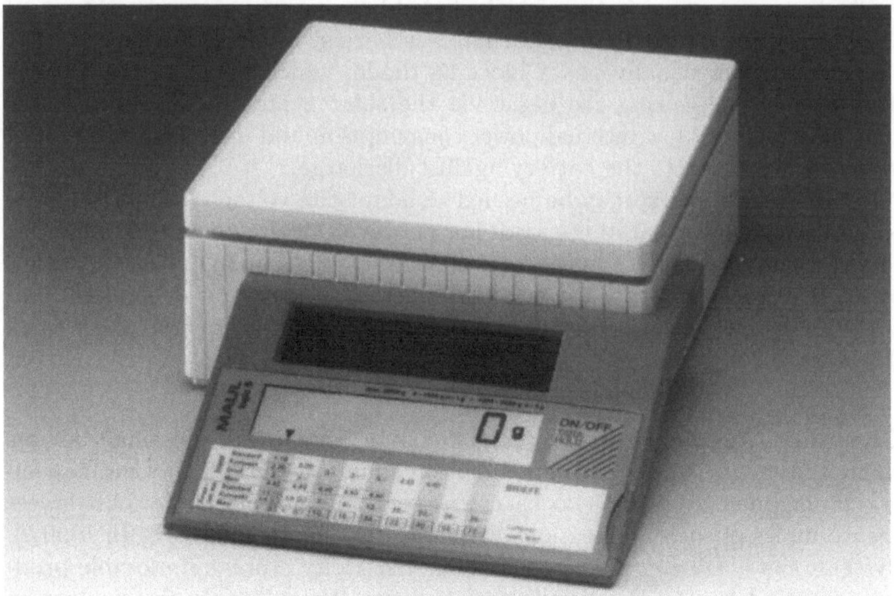

Fig. 2.20. Electric balance for use in households and as letter scale. Source: Maul, Bad König

Another example: fountains can be found in many public parks, squares, cemeteries, and zoos. Usually the water is recirculated rather than being constantly replenished with fresh water. In general, just a few standard solar modules are sufficient to operate the circulation pump. As circulation during the night or overcast periods is not absolutely necessary, a storage element is not needed and the solar generator can be connected directly or via an adaptive converter to the water pump.

Industrial Applications An important field of application for battery-free solar electricity systems are water pumping stations [McNelis 1995]. In this case water, reservoirs serve as product storage devices.

High-power pumps are applied to draw drinking water from great depths for a whole village. The system shown in Fig. 2.21 supplies the drinking water for 4 000 people on Sumba Island in Indonesia. Using a 6 kW_p photovoltaic system, 40 m^3 water can be drawn every day.

Photovoltaically Powered Water Purification Besides water provision, the purification of water also becomes increasingly important [WHO 1993], [UN 1998], [Perodi 1997]. Various technologies exist, like chlorination, UV radiation, and ozonization. From this broad range of purification methodologies,

Fig. 2.21. Photovoltaically powered system to provide drinking water for 4 000 people. Source: TÜV Rheinland, Cologne

the best-suited system must be chosen and integrated in a complete photovoltaically powered system for continuous and safe operation (Fig. 2.22).

Photovoltaically Powered Intelligent Emergency Telephone Emergency telephones along motorways and roads across sparsely populated country are necessary, sometimes even life-saving [Pfanner 1995]. Extensive cabling can be avoided, both during new road construction and when emergency roadside telephones are subsequently installed, if photovoltaically powered emergency telephones with radio contact to a central receiver station are used; see Fig. 2.23 [Steinhülser 2000].

A solar generator of only a few watts power is sufficient, as usually the telephones are not used frequently. The multifunctional emergency call and information system developed at Fraunhofer ISE does not require external cabling, has an autonomous power supply, and operates reliably. In addition to the original emergency call function, contact can be made to various emergency services and information can be obtained. Furthermore, information such as traffic flows and meteorological data can be recorded at the system location and communicated to predetermined central stations.

Photovoltaically Powered Telecommunication Systems Due to their high reliability and low maintenance, photovoltaics are already widely used to power remote telecommunication systems, e.g. repeater stations. Repeaters

Fig. 2.22. Photovoltaically powered water purification unit at a wind-powered water pump in Balde del Sur de Chucuma, Argentina. Source: Fraunhofer ISE, Freiburg

Fig. 2.23. Photovoltaically powered emergency telephone. Source: Fraunhofer ISE, Freiburg

are digital radio signal amplifiers which increase the connection reliability of mobile telephones within radio cells and raise the accessibility level throughout the network. In rural areas and in mountainous regions, repeaters must often be installed operated at sites remote from the public electricity grid; see Fig. 2.24 [Steinhülser 2000].

Because of seasonal fluctuations in solar radiation in temperate zones, an exclusively photovoltaic power supply leads to very large and thus expensive systems. In order to avoid these disadvantages, the photovoltaic generator can be combined with other power generators. Figure 2.25 shows the prototype of a photovoltaic hybrid system, consisting of a PV generator, a battery, and a thermoelectric generator used as auxiliary power source. A microcontrolled energy management system ensures fully-automatic operation with minimal consumption of fossil fuel and appropriate battery operation management.

2.3.3 Remote Habitation

Solar Lantern for Rural Households in Developing Countries A photovoltaic application which is particularly interesting with regard to its ecological and social aspects is the replacement of fossil-fueled light sources – usually petroleum lamps – in remote regions of developing countries with photovoltaically powered lights.

The robust lighting unit illustrated in Fig. 2.26 consists of the light, an electronic ballast, the rechargeable battery, and a charge controller. During the day, it is connected via a cable and plug to the permanently mounted solar generator for charging. In the evening it can be used as a portable light inside or outside the dwelling and can be carried and positioned wherever it is needed most. A solar generator with $4\,W_p$ (module size $(20 \times 20)\,cm^2$) is sufficient for a daily lighting period of 3 to 5.5 hours (depending on the solar radiation). At the same time, the solar-powered light, with a luminous flux of about 80 lm, is three times as bright as a petroleum lamp.

Solar Home Systems The designation “Solar Home System” has become a well-defined concept [Asia 1996, Njaimi 1992, Perodi 1995]. A 50 W photovoltaic module (size $50 \times 100\,cm^2$) typically supplies power for three lamps and a black-and-white television in a single household; Figs. 2.27, 2.28. A lead-acid battery with a charge controller stores the energy from the day for the night and can run alone for over two or three overcast days. Depending on the size of the local market, customs duties, taxes, and the share of locally manufactured components, the cost of a Solar Home System is between $500 and $1 500 (US).

In many situations, Solar Home Systems offer cost advantages over the classic alternative for electrification, which is the extension of electricity grids into sparsely populated regions and connection of power lines to all rural households [Preiser 1995]. The World Bank assesses the average cost worldwide of grid extension to be $900 (US) per household. In unfavorable cases,

Fig. 2.24. PV powered repeater station on a hill in Tuscany, Italy. Source: Fraunhofer ISE, Freiburg

Fig. 2.25. Prototype of a hybrid system (photovoltaics and thermoelectric converter) to power a remote repeater station. Source: Fraunhofer ISE, Freiburg

with long distances, difficult terrain, and low population density, a power line for one household can cost many tens of thousands of dollars.

The investment needed for this cannot be financed via the low energy consumption of rural households, which ranges between fractions of kilowatt-hours and a few kilowatt-hours per day. Political decisions to electrify rural areas despite this imbalance have contributed to electricity utilities going into debt in developing and threshold countries, so that necessary investments to modernize distribution grids and power stations cannot be financed. The result is the poor quality of the electric power supply, which can be observed in many regions around the world. Increased privatization of the power supply is intended to provide a remedy to the described situation [Foley 1990, Foley 1995].

The situation described above is the main reason why small diesel generators in the power range of some kW are installed in many remote locations. Because of their low investment costs, they appear economically attractive. However, calculating the lifetime cost shows that the efforts needed for repair and maintenance and the fuel costs add up to considerably high sums.

Decentralized electrification with privately financed Solar Home Systems fits in perfectly with the aim of sustainable and cost-effective rural electrification. It is therefore actively supported by the energy planning authorities of

Fig. 2.26. Photovoltaically powered light in a hut in Tunisia. Source: Ludwig-Bölkow-Stiftung, Ottobrunn

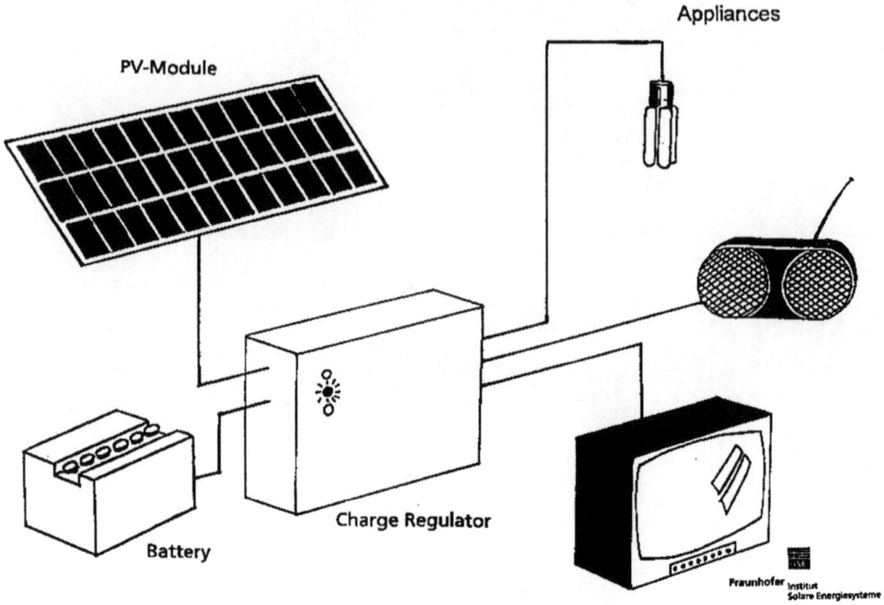

Fig. 2.27. Block diagram of a Solar Home System to cover basic electricity demand (lighting and communication) in rural households

various countries and by multilateral organizations such as the World Bank [The World Bank 1996].

Light, which allows information and communication for private users and lengthens the working time into the evening in farms and shops, schools, and community centers, cannot be underestimated in its importance. Solar Home Systems can provide this light [Perodi 1998]. Electricity also solves other supply problems. Workshops, food storage and processing, infrastructure, tasks in health services and administration, and the development of tourism demand greater daily amounts of energy and power than can be supplied by a Solar Home System. The next electrification stage consists of larger individual systems and then local networks (see below).

PV-Powered Battery Charging Stations Experience with locally centralized battery charging stations (Fig. 2.29) is by far not as convincing worldwide as it is with Solar Home Systems. Normally there is no deep discharge protection of the battery foreseen, which besides other effects causes sulfatation of the electrodes and thus quickly curtails the lifetime of the battery. Using the batteries in a full cycling process means that is it not the corrosion coefficient that determines the lifetime, but the cycle capability. Therefore battery lifetime tends to be very short, only some hundred cycles, especially with locally produced car batteries [Wiengmoon 1997].

Fig. 2.28. Sukatani, the first SHS pilot village in Indonesia, was the origin of the currently running rural electrification program to electrify one million rural households in Indonesia. Source: Fraunhofer ISE, Freiburg

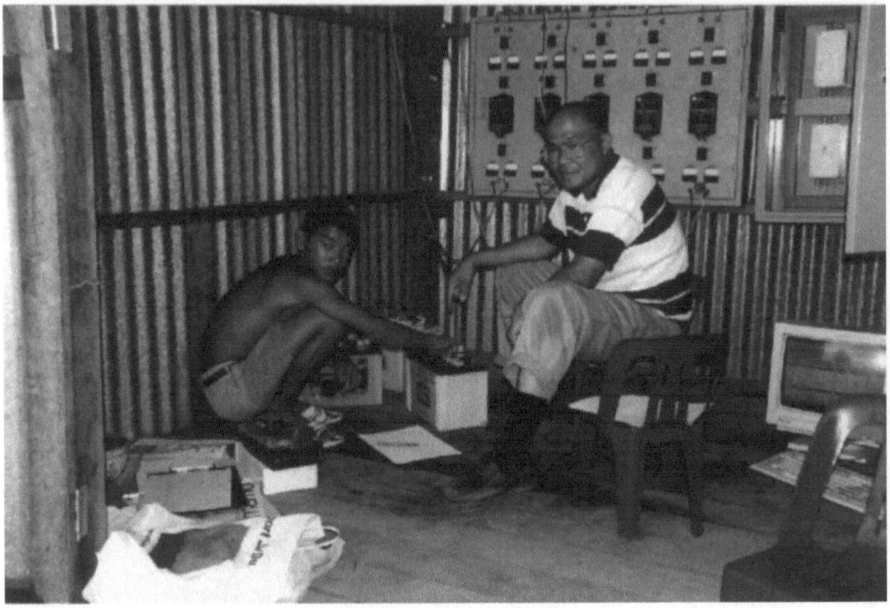

Fig. 2.29. A typical PV-powered battery charging station in rural Thailand. Source: Fraunhofer ISE, Freiburg, [Wiengmoon 1997]

Another aspect often neglected with battery charging stations is that people have to carry dangerous freight over quite a long distance. In industrialized countries no one would expect this of somebody, even if the means of transport are more comfortable, as they are in rural areas of developing countries. The danger of accidents when transporting batteries, the high risk of wrong connections afterwards, and very poor charge and discharge protection regimes yield unsatisfactory operation of this kind of electricity service.

Village Power Supply Systems While Solar Home Systems for the decentralized power supply of rural households are considered to be a standard system ready for the market, and while the suitability of PV hybrid systems to power single houses has been proven in various pilot and demonstration programs all over the world, this positive experience is up to now not available for central power stations to supply remote villages with off-grid PV power. Nevertheless, research and development in rural electrification are increasingly directed towards these systems. The main reason is that with a central system, installation and operation costs may be less than the single house supply option. Today, as more and more electric utilities enter the off-grid market, their interest is to lower the cost as much as possible for operation, maintenance, money collection etc. needed to assure robust energy service. Experience and reports from those involved in PV dissemination programs

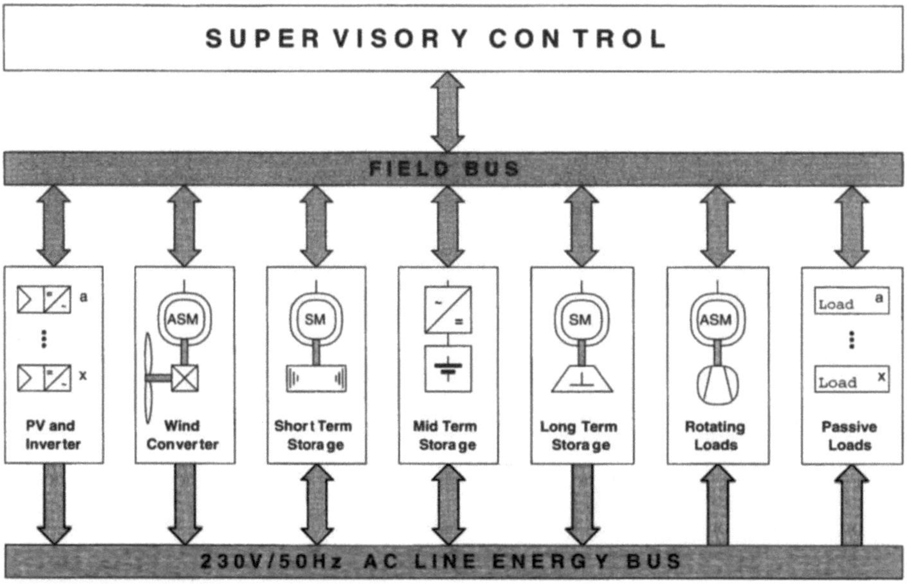

Fig. 2.30. Concept for a village power supply system, including various electricity generators and energy storage components. The whole system is managed by a control unit implemented in the central inverter. This concept which relies on an AC bus to whitch all power generators, storage components, and loads are connected, was developed by ISET Kassel in the early nineties [Kleinkauf 1994], [Kleinkauf 1996]. Source: ISET, Kassel

for rural electrification confirm that technical problems in the village supply are not the main barrier preventing lasting success. Rather, an appropriate introduction method, user involvement at various levels, and planning which allows flexible reaction to the situation as it develops are necessary to achieve optimal shared use of the limited resource. Today there are technical components, like the inverter, the charge controller, and energy management systems, available in various countries that are much better suited now than they were in the early demonstration projects. Figure 2.30 shows a typical configuration of a village power supply system. In this system design, AC coupling is demonstrated, in whitch all components are connected via an AC bus and the central inverter manages all the energy flows.

R&D efforts in the development of village power supply systems are now focused on the fair distribution of the energy available, and a suitable limitation of the power consumed by each single household. Prepayment schemes or energy limiters are under development and have been demonstrated by several companies and international pilot programs. These systems meter system energy consumption in each household and cut the load if preset values or the energy budget have been consumed. A typical representative is

Fig. 2.31. PV-Diesel-Hybrid system in Salopangkang/Sulawesi. Source: Fraunhofer ISE, Freiburg

TRANSINDO, a French-Indonesian project supplying power to four remote villages in Indonesia, each consisting of 300 to 400 households; see Fig. 2.31.

Photovoltaic Hybrid Systems If higher power is needed or if conventional household appliances and industrial equipment are to be used, a system voltage of 230 VAC is desirable.

The usual supply levels, classified according to small and medium loads, and local, regional, and national supply with the corresponding voltage levels and power ranges, are illustrated in Fig. 2.32. For very low power requirements, pure photovoltaic systems with an appropriate energy storage unit are the best option. With increasing power demands, other electricity generators, like wind turbines and solar thermal generators, become technically and economically attractive. Small standalone grids that are supplied by various electricity generators are the next step in remote area power supply. With increasing demand, connection to the public electricity grids makes good sense. Here, of course, all renewable energy technologies can play a role; a decision on their application, however, must consider the energy prices of conventional power stations driven by fossil or nuclear fuels. Due to price reductions in recent years, photovoltaic power technology is now cautiously expanding into the "supply of small and medium loads" range.

To increase the reliability of off-grid power supply systems and to reduce investment costs for any particular energy service, larger systems (i.e.

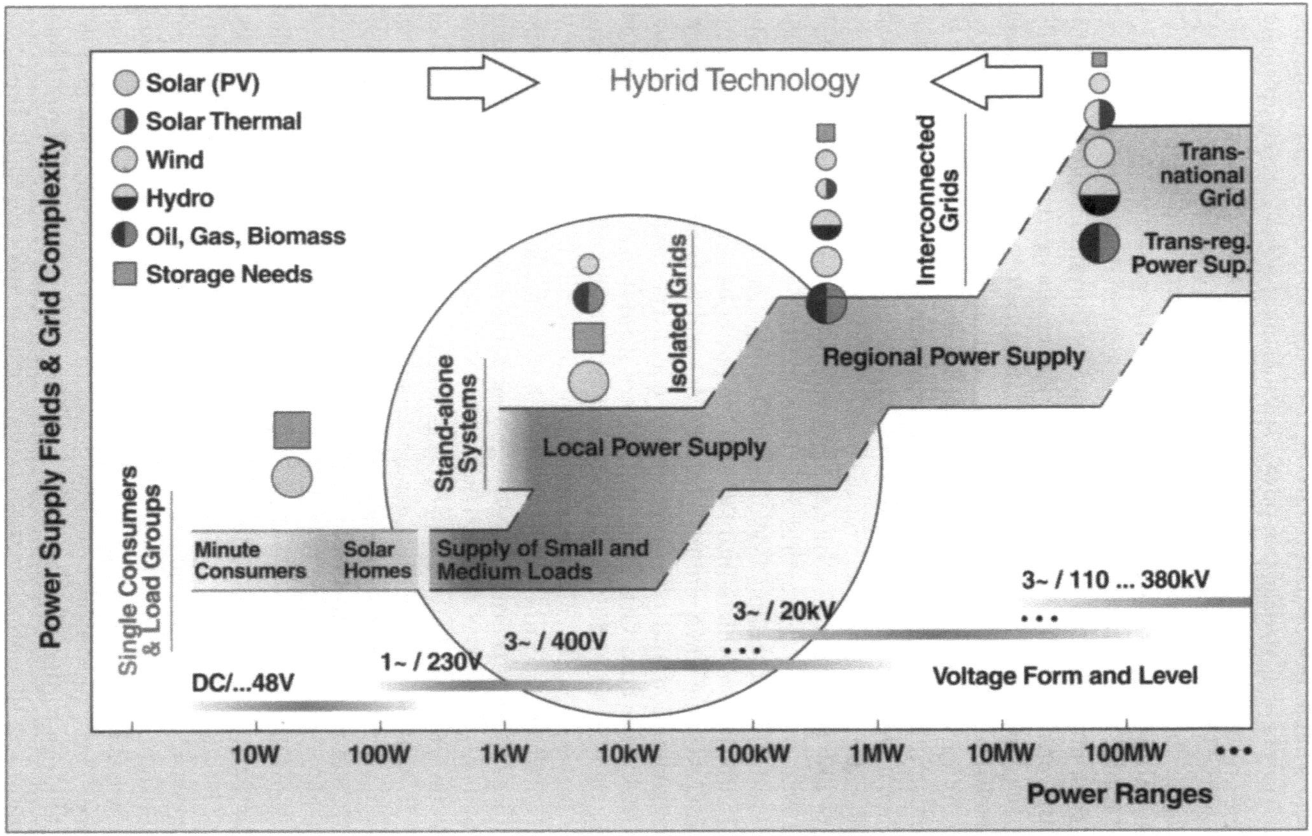

Fig. 2.32. Classification of power supply technology for electrification with renewable energy – energy sources, application classes trends. Source: ISET, Kassel [Kleinkauf 1998]

those that provive several kilowatt-hours a day), are typically constructed as hybrid systems: a diesel generator and a photovoltaic generator. Depending on the availability of natural resources, other electricity generators like wind converters or hydroelectric power stations may be used to complement the power supply system. A battery bank ensures that power is available continuously. The systems are normally designed to provide conventional AC electricity at the voltage level and frequency that is standard in the respective countries. Under favorable weather conditions, the consumer's total energy demand is met by the solar generator which operates silently and without exhaust gases, an environmentally friendly option. Surplus energy is stored in batteries. During the night or unfavorable weather, energy demand is initially met by the batteries. If there is danger of deep battery discharge, a diesel or gas-fueled generator provides electricity and simultaneously charges the battery. At windy sites, the system can include a wind energy converter. As the photovoltaic generator and a wind-energy converter largely complement each other if they are dimensioned correctly, the operating time of the fossil fueld generator, and thus fuel consumption, are reduced.

There are two main design concepts competing in the current market [Bopp 1997]: most of the hybrid PV-diesel-battery systems realized worldwide are so-called DC-coupled systems. Different generators work via separate charge controllers/rectifiers into DC storage batteries. An inverter generates the desired AC sine wave and matches the AC grid in frequency and voltage. If power generation and consumption are equal, the battery does not take part in the power flow. Otherwise the battery stores the surplus of generated power or delivers the additional power needed by the load. In order to protect the battery against damage, one charge controller, in most cases the PV charge controller, disconnects the inverter before the battery becomes completely discharged. If there is not enough renewable power and if, at the same time, the charge state of the battery is low or consumers need more power than the inverter can deliver, a backup generator is started from the control system. In this case the backup generator supplies consumers directly with AC power and charges the battery via its own rectifier. The big advantage of the DC-coupled system layout is that it is proven in many applications, and reliable components are available on the market. The need for adaptive charge controllers for each generator, however, can lead to systems with relatively high installation costs, especially if there are more than two generators. Even in Germany there are still many houses, hikers' inns, and mountaineering lodges which are not connected to the public grid because of the long distances and the resulting connection costs. The only option open to owners up to now has been to install their own electricity generator – usually a diesel generator. These generators are mostly used to supply the consumer load directly. As a result, the diesel generators are often operated in the partial load range, and when they are turned off, no electricity is available at all. However, continuous operation is undesirable, not only because of the low partial load

Fig. 2.33. Rotwandhaus with its solar generator and wind energy converter. Source: Fraunhofer ISE, Freiburg

efficiency, noise, and exhaust gases, but also because of the limited lifetime of combustion motors.

The Rotwandhaus, a hikers' inn which is open throughout the year, was equipped with a hybrid system consisting of a photovoltaic generator, a wind energy converter, and a diesel generator, in the course of necessary renovation work (Fig. 2.33). The roof-mounted solar modules convert sunlight to electricity silently, without polluting the environment, and without exhaust gases. Electrical appliances are supplied directly and the surplus is stored in a battery while there is sufficient sunlight; power is drawn from the battery at night or during overcast weather.

When the wind is strong enough, the wind energy converter supplies power and charges the battery. As soon as the battery is fully charged, the wind energy converter is throttled when the wind is too strong. The inverter converts direct current from the battery to 230 V alternating current, so that all conventional electrical appliances can be operated. The necessary supply reliability, even during extremely unfavorable weather conditions, is guaranteed by the diesel generator. A computer monitors and controls the whole system to ensure that the supplied energy is used as effectively as possible.

The other possibility for hybrid system design is the so-called AC-coupled system (Fig. 2.30): The heart of an AC-coupled system is a bidirectional inverter. The photovoltaic modules and all generators are connected to an AC bus bar via suitable inverters. Thus, AC consumers are primarily supplied with electricity and only surplus power is fed back to storage batteries. Power management is done by only one electronic device, the bidirectional inverter. This can lead to a simplified system design. The inverter now must be able to work in different modes (charging or discharging the battery). It must also define the local AC grid in frequency and voltage, and it must be synchronized to any other synchronous electric generator in the system. Due to these stringent requirements for the inverter, there are only very few bidirectional inverters available on the marked, some of which are still in prototype form.

Battery Storage Energy storage is a key element in solar energy supply. Thirty percent of the lifetime costs of solar off-grid systems, or even more, may be attributed to storage. Although a variety of storage technologies is under development, the lead-acid battery still is, and will be for some years to come, the workhorse of electricity supplies in remote areas. There are a number of technical solutions to the problem of energy storage. The energy stored in the electric field of a capacitor is widely used for storage times of microseconds, as it smoothes the output voltage of a PV generator which feeds a DC/AC inverter directly. Special capacitors with high energy density are used in very small systems with energy consumption no greater than mWh per day (Supercaps).

The most promising storage technologies today for the application discussed here are electrochemical systems [Linden 1995]. One possibility is to split water into hydrogen and oxygen, using an electric current in an electrolyzer. The gases are stored and reconverted in a fuel cell when electric power is needed. The hydrogen can also be burned, and the heat converted to power in a heat engine, a thermoelectric [Roth 1996] or thermophotovoltaic [Schubnell 1997] generator. Electrolyzer/fuel cell systems have been demonstrated, but the technology still has to go through a long process of maturation before it reaches the market. The auto industry is spending large sums of money to develop a cheap and reliable fuel cell for electric vehicles, which of course might eventually be used as a storage component in solar systems.

The number of electrochemical energy storage systems which have found their way into widespread application is large. Li-ion, NiCd, NiFe, NiMH, NiZn, Zn/AgO, and last but not least Pb/PbO2 accumulators are available as reliable products. Of these seven systems, only one, the Pb/PbO2 (lead-acid) battery, is presently suited to rural energy applications. The other batteries mentioned are ruled out either by their extraordinarily high price or their low storage efficiency.

The following example underlines the key role which battery storage has to fulfill in hybrid systems. The Unterkrummenhof is a typical hybrid system, with a 4.5 kW_p PV generator, a diesel generator, and 32 kWh of lead-acid battery storage. It has been in operation since 1992 in the Black Forest in Germany [Gabler 1995]. It must deliver 10 kWh per day to the Unterkrum-

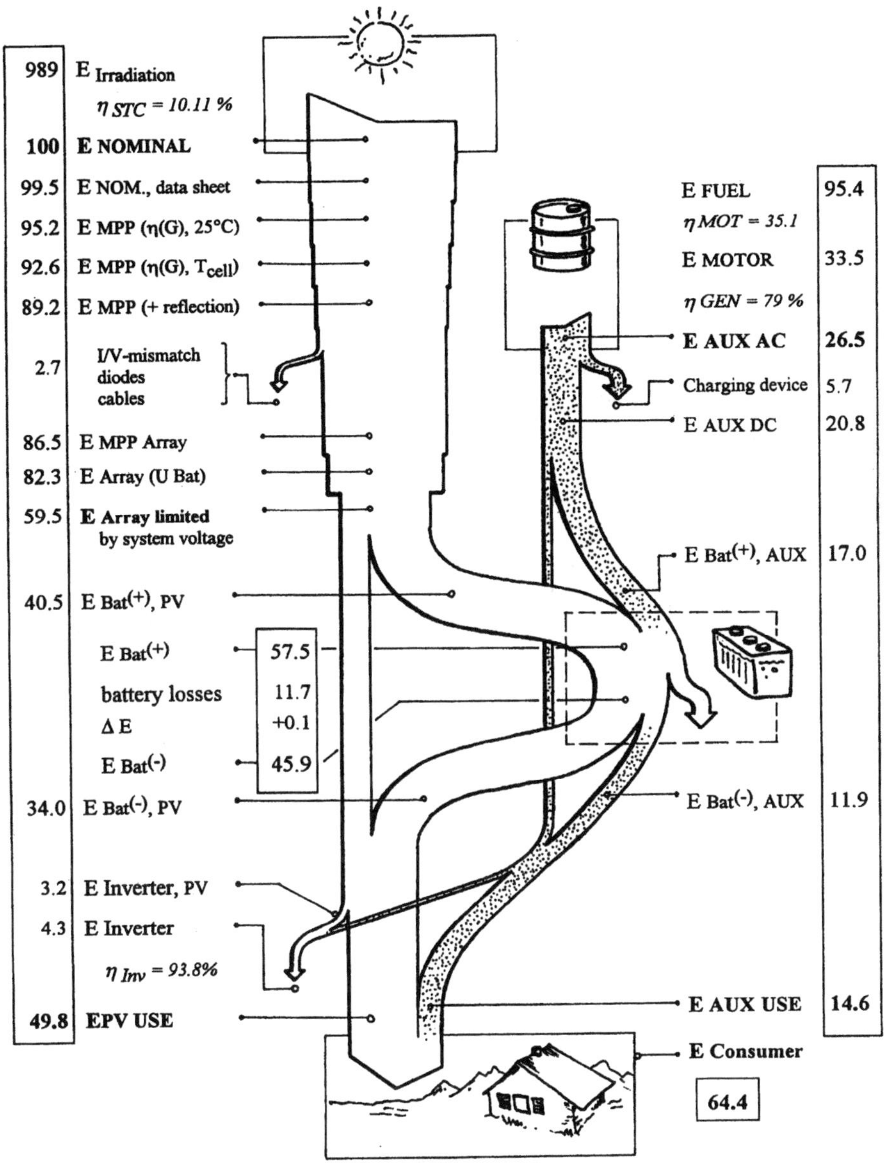

Fig. 2.34. Energy flow diagram of the Unterkrummenhof autonomous energy supply system (PV-diesel-battery) near Freiburg, Germany. As can be seen, a considerable share of the energy flow must come from the storage battery

menhof hikers' inn. Modeling calculations based on measured data show the flow of energy in the system and the effect of internal losses (Fig. 2.34). Two thirds of the power delivered (Fig. 2.34: E consumer) is drawn from the battery storage. The numbers given in the diagram are normalized to the nominal energy production of the PV generator under the radiation conditions at the site.

As battery lifetime is one of the key factors that determine lifetime costs of the whole standalone photovoltaic system, one must observe a complex set of rules when aiming for maximum battery lifetime, by defining suitable end-of-charge and end-of-discharge thresholds, avoiding deep discharge, avoiding acid stratification in the electrolyte, avoiding high battery temperatures, assuring frequent full charging, avoiding aberrant single cells in a series of connected batteries, etc. [Bopp 1995, Sauer 1997]. Battery manufacturers normally give rules for the allowed operation conditions of their products. These rules can be derived from the rules given above. In photovoltaic standalone systems it may be difficult to follow these rules at all times. A compromise must be found between battery requirements for regular recharging and in investment in expensive photovoltaic generators, so energy may not be available during low insolation periods to recharge the battery. The availability of energy for the user may at certain times be so important that deep discharge of the battery is permitted. Or, and this may be the most important source of rule violation, the system manufacturer, installer, or user may not know, or at times may not want to know, how the battery should be treated. To protect the battery as well as possible from harmful operating conditions, electronic equipment controls energy flow into the battery or inside the battery bank. Experience shows that battery lifetime in well-operated standalone systems can over seven to eight years, sometimes even more, a considerable improvement over earlier experience, where in some applications battery lifetimes were only about 3–4 years.

Balance Systems Besides electricity generators and energy storage, other components, so-called balance systems are necessary to form a well-behaved standalone system. Energy management units play a key role: the charge controller, inverter, DC/DC converter, and energy management systems. As discussed in the chapter on hybrid systems, auxiliary power supplies are a good option to support redundancy in a standalone system, and thus to increase power-supply reliability and in many cases reduce overall electricity costs.

Charge Controller and Energy Management Systems Although photovoltaic systems can be used without charge controllers, and this practice can very often be found in small PV systems, it must be stated that long-term operation of a standalone photovoltaic system required that one avoid overcharge and deep discharge. As seen in the previous chapter, battery costs over the lifetime of a PV system take the major share. Battery lifetime again depends to a large degree on operation strategy. In Table 2.1 the most important rules to achieve long battery lifetime are summarized.

Table 2.1. Rules for battery management and their positive effects when they are applied

Rule for battery management	Effect when applied
Avoid high voltages while charging the battery	Less corrosion and loss of water
Avoid low voltages during discharge of the battery	Less corrosion
Avoid extended periods without full charging of the battery	Less sulphatation
Avoid deep discharge	Less sulphatation and growth of dendrites
Avoid reverse charging of the cells	Less degradation of negative electrodes
Avoid stratification of the electrolyte	Less sulphatation
Avoid high battery temperatures	Slower aging processes
Overcharge the battery slightly once a month	Frequent full charging is assured

It must be pointed out that some of the rules are generally inconsistent to each other (e.g. full charging needs high voltages but high voltages accelerate corrosion). In photovoltaic standalone systems it may be difficult to follow these rules at all times. A compromise must be found between the demands of the battery for regular recharging and the investment in expensive photovoltaic generators, so energy may not be available during low insolation periods for recharging the battery. As the application of the rules is the responsibility of the charge controller it becomes evident how complex the technical design for this important electronic system component can be.

When system complexity increases, other aspects must be considered in designing a standalone PV system: with appropriate energy management schemes, the use of the electricity generated and the lifetime of sensitive system components can be further optimized, e.g. turning off lower-priority loads, turning on water pumps to access energy to fill water storage, starting the backup generator to avoid critical situations for the battery. These energy management units must be designed specifically for designated application. In systems where human users play a role, it is highly recommended that the system display units inform the user about the current status of the system and give them advice about how to react in an appropriate way. Only when the user participates in the whole process of system management will the user operate the system responsibly [Butera 1989].

Inverters Inverters are power electronic devices used in various photovoltaic (PV) system configurations:

- grid-connected systems
- standalone systems with rechargeable batteries
- pumping systems without storage batteries

Inverters for Grid-Connected Systems The planning of a grid-connected photovoltaic system begins with the choice of a suitable inverter [Wilk 1992]. This determines the system voltage on the DC side, and the solar generator can then be configured according to the input characteristics of the solar inverter. The inverter is the second most important component of a grid-connected photovoltaic system (after the solar generator). Its task is to convert the direct current generated by the solar cells to 50 Hz alternating current conforming to the grid. In contrast to inverters intended only for standalone operation, those intended for parallel operation must respond just as well to the grid characteristics as to the solar generator performance. As all of the solar current flows through the inverter, its properties fundamentally affect the behavior and operating results of the photovoltaic system. Apart from the efficient conversion of direct to alternating current, the inverter electronics also include components which are responsible for the daily operation mode. They ensure that operation starts at the right time in the morning, as soon as the solar cells deliver enough power. Unsuccessful start attempts require energy from the grid and should be avoided by good controls. During the day, the optimum working point on the I-V characteristic curve shifts according to fluctuations in solar radiation and module temperature. Intelligent inverter control includes maximum power point (MPP) tracking and continuous readjustment to the most favorable working point. Protective devices are also integrated into the inverter, which automatically disconnect the system if irregularities in the grid or the solar generator occur. Today most inverter models are additionally equipped with data loggers and measurement computers, which allow the power, voltage, current, and other operating parameters to be recorded continuously. These data can be read out at intervals via a serial interface with a laptop computer and analyzed.

Inverters for Stand-Alone Systems Due to the specific operating conditions of standalone inverters, various design aspects must be considered. In a typical domestic power supply, the ratio of peak power to average power is about 25:1, so the inverter must have a high efficiency of approximately 90 %, particularly in the partial load range (5–10 % of the rated power). Only a few inverters satisfy this condition, together with sinusoidal output voltage and the capacity to withstand short-term, double to triple overloading. Depending on the requirements, both sinusoidal and rectangular waveforms can be used.

The most important requirements on inverters for standalone photovoltaic systems are summarized in the following list.

- Large input voltage range (−10 % to +30 % of the rated voltage)
- Output voltage as close to sinusoidal as possible
- Little fluctuation in the output voltage and frequency
- ±8 % voltage constancy, ±2 % frequency constancy
- High efficiency under partial loading; an efficiency of at least 90 % at 10 % partial load
- Ability to withstand short-term overloading for appliance starting conditions, e.g. two to three times the rated current for 5 s for refrigerators and washing machines
- Lowest possible overvoltages for inductive and capacitive loads
- Half-wave operation possible, caused by power reduction with a diode, e.g. in hair dryers
- Able to withstand short circuits

Auxiliary Generators In general, diesel generators are used nowaday in larger photovoltaic hybrid systems with an energy demand of around 10 kWh/day. For systems with a daily energy demand of a few hundred Wh up to a few kWh, small combustion motors, small fuel cells, thermoelectric generators (TEG), thermophotovoltaic generators (TPV), or appropriate thermodynamic converters such as small Stirling motors all come into question, in principle. If adequate commercial availability is regarded, at present the choice is narrowed essentially to small combustion motors (diesel, petrol) with a coupled generator, and thermoelectric generators. While the investment costs and the availability in nearly all countries make petrol and diesel generators very attractive, the inherently high reliability, low maintenance, and the almost silent operation and simple remote starting facilities suggest the application of thermoelectric generators in small photovoltaic hybrid systems (Table 2.2).

A significant disadvantage of combustion motors is their need for regular maintenance. Car manufacturers recommend for diesel motors a change of lubricant at least every 10 000 km. If we assume an average speed of 50 km/h this means that complete service is necessary after 200 hours of operation. Keeping in mind thatback-up generator in a standalone system is working in optimized systems at least for 2 hours and in conventional systems about 6 to 8 hours a day, this means that in optimized systems every three months and in conventional systems every 30 days, a complete maintenance service has to be calculated. Together with the high costs for fuel transport, this is the main reason for the low attractiveness of combustion motors in off-grid electrification.

Future Developments in Off-Grid Power Supply with Photovoltaics Rural electrification with photovoltaics has become a rapidly emerging market in the last few years. While grid extension is getting more and more difficult due to the high costs of connecting remote households, with their

Table 2.2. Comparison of the characteristics of small, commercially available auxiliary generators [Steinhülser 2000]

	Small combustion motors		Thermoelectric
	petrol	diesel	
Electric power range	> 0.3 kW	> 3 kW	> 0.03 kW
Efficiency value at full load	app. 5–15 %	app. 20–25 %	app. 3 %
Adjustability	good	good	limited
Remote starting	yes, > 3 kW	yes	yes
Reliability	medium	medium	high
Maintenance demand	high	high	low
Environmental acceptability	pollution emission CO_2 emission noisy	pollution emission CO_2 emission noisy	pollution emission CO_2 emission almost silent
Commercial manufacturers available	globally	globally	few
Fuel:			
Type	petrol	Diesel	propane/butane, etc.
Availability	medium	high	medium
Consumption	0.6–1 l/kWh	0.4–0.5 l/kWh	2–2.5 kg/kWh

relatively low energy consumption for lighting and communication, about two billion people worldwide remain without access to electricity as the modern form of energy supply. As the population, especially in rural areas of developing countries, grows rapidly, there are more people each day without access to electricity than the day before! This is the main reason why rural electrification programs have been initiated in many countries. These programs explicitly foresee the application of photovoltaic systems to provide electric power to remote rural households and public facilities, e.g. in Argentina, 300 000 households and 6 000 public facilities, in Marocco all villages in rural areas shall be electrified by the year 2010, whitch takes into consideration that about 200 000 households must be supplied by photovoltaics. In many other countries like China, Indonesia, India, Sri Lanka, Philippines, Mexico, Bolivia, Kenya, Uganda, South Africa, large programs do exist or are in preparation for realization in the near future. The World Bank, as one of the major support organisations in rural electrification, is involved in more or less all these national programs, thus helping to cover the high upfront costs of PV rural electrification and establishing innovative financing schemes. Most of these programs today concentrate on Solar Home Systems, as these systems are relatively easy to install and operate. However the key to success or failure of these programs lies in the establishment of appropriate infrastructures for installation, operation, and maintenance of these systems, as well as in the introduction of schemes for quality certification, microfinancing, and adequate involvement of users [Schweizer 1995, Preiser 1995, Gabler 1998].

After some disastrous experiences with village power supply systems in the seventies and early eighties, they see today a revival in many rural electrification programs. Reasons for this are that much better suited and more reliable components are available today and that more and more utilities or other energy service companies (ESCO) are starting to enter the market. With these new players a different concept appeared, which is the provision of energy service instead of selling PV systems. In Argentina, for example, companies about to engage in rural electrification – and benefiting from the subsidy scheme set up there – must sign a contract to serve the rural population with electricity for at least 45 (!) years. Here village power supply systems are often applied: a small electric grid is created, which may be considered similar to the existing public electricity grid. However, due to the strong limitation of the resources compared to a large electricity grid, new models have to be developed to plan, market, operate, and sell this service. The involvement of users in these central systems is seen as essential, as the behavior of the whole community does have a major impact on the appropriate use of the limited power and energy provided. Appropriate financing schemes, like pre payment meters or the involvement of microfinancing institutions in the whole process, are new developments that have to be further elaborated in the near future.

Today about two billion portable appliances are used worldwide, and the market is rapidly growing. In the near future, this number is expected to double. Watches, laptops, mobile phones, PDAs, etc. are characteristic of the new information society. This means the future of information is free of cables. Most of these devices are powered by batteries, but they quickly go dead and have to be recharged at a grid-based charging station. Photovoltaics are considered to be one of the best solutions to cover the typically low energy demands of these grid-independent devices. First prototypes do exist. The development of the market will be successful if adapted systems can be developed that can cope with the often extreme restrictions of miniaturization and a frequent lack of sunlight or at least of artificial light. Highly efficient solar cell, highly miniaturized and efficient energy conversion, and new product designs adapted and suited to the energy source, will be the key factors in successful market penetration.

In parallel with the power supply of the appliances themselves, a need arises for repeaters of the signals to be transmitted that will be lower in their energy and power demands but will have to guarantee permanent availability. As these repeaters will have to be located on peaks or tops of hills and mountains, grid connections will very often bee too expensive. Therefore low-maintenance hybrid systems will gain a natural market, where they can compete with extremely stringent requirements for reliability and ease of operation.

In all standalone applications, the storage unit is one of the technical components which still needs the most development. Often described as the weakest link in the chain, it is responsible for a large share of the costs in long-term operation. Thus optimization of the interaction of all components in the system with the aim of reducing the overall lifetime costs is today the main issue this sector. As the battery industry has also recognized, there is a great need for the development of better battery management principles, adaptation of the existing battery technologies to the charging and discharging requirements in photovoltaic systems, and the development of completely new and better-suited storage units.

2.3.4 Grid-Connected Photovoltaic Systems

Decentralised Grid-Connected Photovoltaic Systems Photovoltaic systems can also be connected to the public electricity grid via a suitable inverter (Fig. 2.35). Energy storage is not necessary in this case. On sunny days, the solar generator provides power for the electrical appliances in a house (Figs. 2.36 and 2.37). Excess energy is supplied to the public grid. During the night and overcast days, the house draws its power from the grid [Erge 1998, Heydenreich 1996].

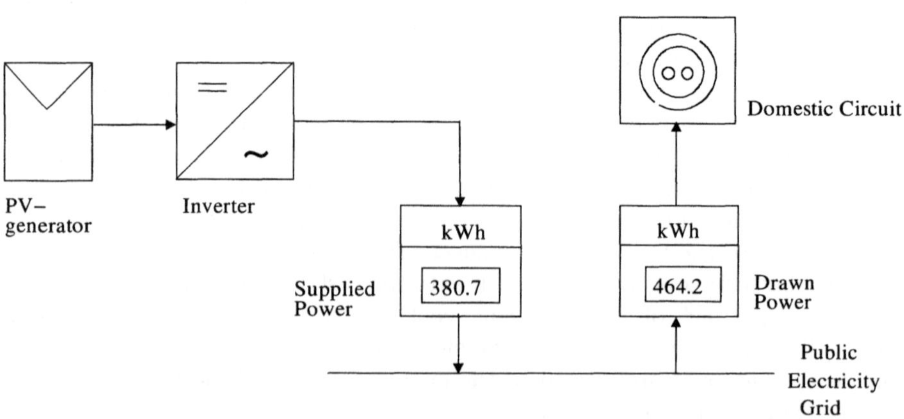

Fig. 2.35. Block diagram of a grid-connected photovoltaic generator. The location of the two energy meters may vary, depending on the adopted payment scheme

Fig. 2.36. Grid-connected photovoltaic roof on a private house in Germany. Source: Fraunhofer ISE

Photovoltaic systems operating in parallel with the grid have great technological potential [Erge 2000]. However, without subsidies from government or utilities, they are not yet financially competitive. In the last few years, however, in various countries programs have been initiated that encourage the dissemination of grid-connected systems, e.g. the 70 000 Photovoltaic Roofs Program in Japan, or the 100 000 PV Roofs Program in Germany. In addition, national laws which require paying higher fees for renewable electricity fed into the grid, like in Germany 99 Pfennig per kWh produced from photovoltaics, makes it more and more attractive for private consumers and investors to become engaged in this technology.

Central Grid-Connected Photovoltaic Systems While the photovoltaic systems discussed above are usually small decentralized systems in the lower kW range, it is also possible to build large central photovoltaic power stations in the upper kW to MW range. It is then possible to feed directly into the medium- or high-voltage grid (Fig. 2.38).

Joint Ownership Sixty percent of the total population in Germany live in rented apartments. Consequently, they will generally not be able to install a grid-connected PV-plant on the rooftops of their house. When these people are interested in PV, they can only become PV owners if PV projects are established with joint ownership. To create this joint ownership, a legal entity

Fig. 2.37. Facade integration on the new train station in Freiburg, Germany. Source: Fraunhofer ISE

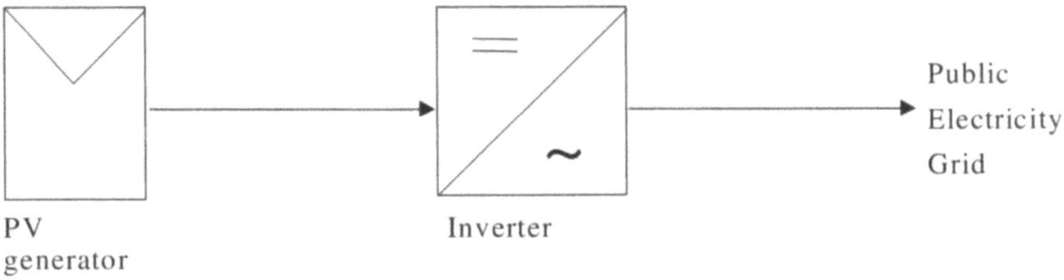

Fig. 2.38. Block diagram of a grid-connected photovoltaic power station

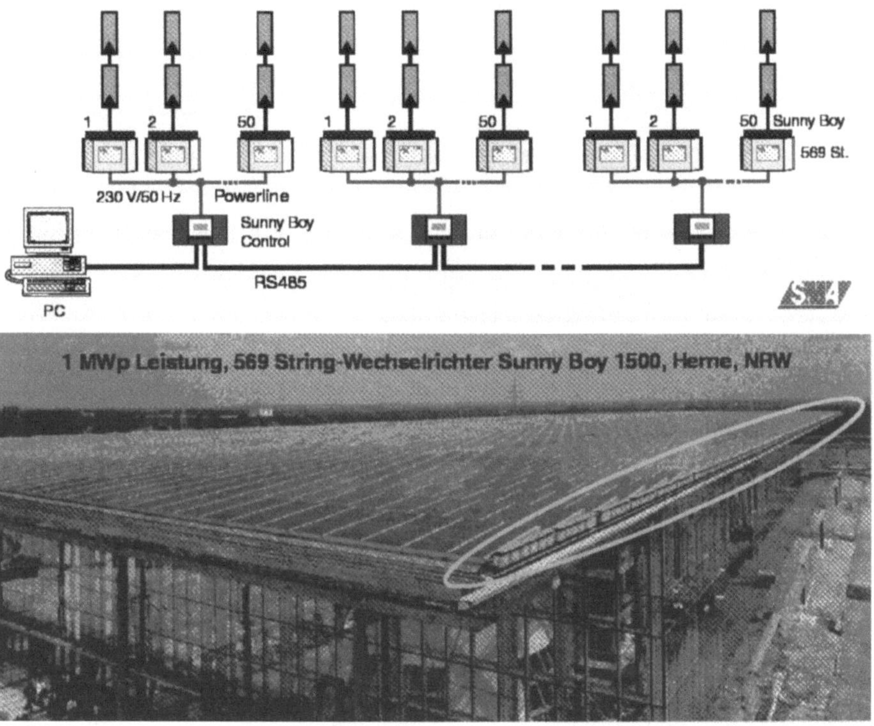

Fig. 2.39. Central PV system installed at Mont Cenis, Herne, Germany. Source: ISET Kassel

is setup to deal with project management, as well as with the related financial questions and the distribution of profits generated through investment. In most cases, contracts are signed between various project partners to resolve management issues during project implementation (see Fig. 2.40).

Contracts contain operational, financial and legal items concerning the collaboration of the different parties in project implementation. Signing parties are mostly the project developer, the utility, the end-owner of the PV system, and the so-called limited partners. The projects are either financed by

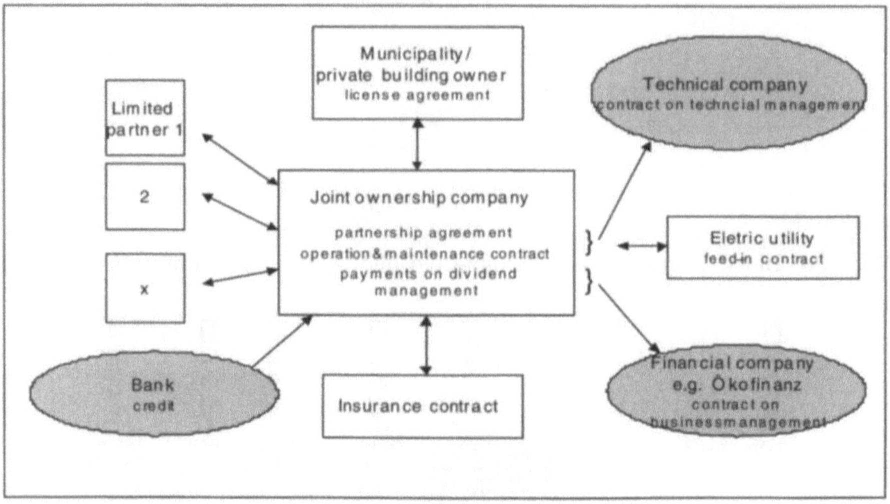

Fig. 2.40. Example of the organizational structure of a joint ownership project [Hille 2000]

Fig. 2.41. Joint ownership PV installation of 100 kW on the soccer stadium in Freiburg, Germany [Hille 2000]

purely equity capital of the owners, or a mix of equity and debt capital through credits. A (major) portion of capital is generated by the project through a public placement. In this case, transaction costs are minimal, and risk funding is supplied by the larger public participating with risk equity.

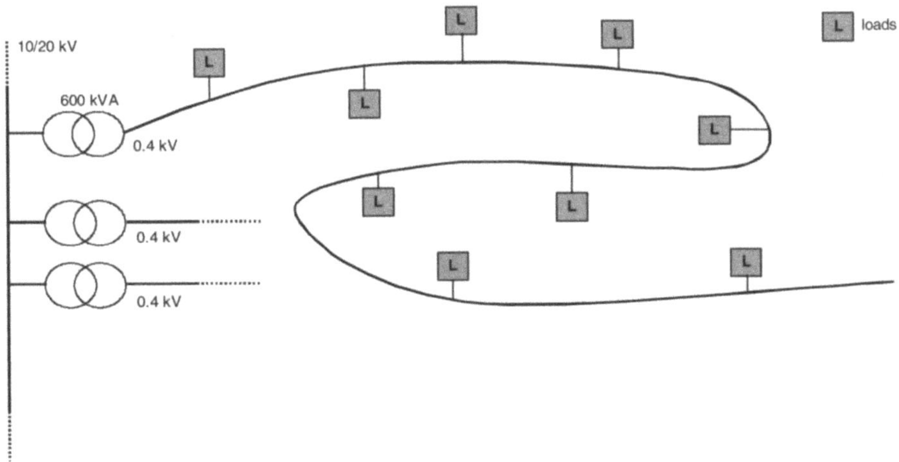

Fig. 2.42. Present structure of electricity distribution grids

Distributed Power Generation Today's electricity supply systems in industrialized countries are based on large central power stations that supply consumers via a dense distribution grid (Fig. 2.42).

This structure will suffer increasingly in liberalized markets, as it is not flexible enough in the choice of primary energy sources to allow energy flows and energy costs to be optimized. In addition, obtaining authorization for large technological power plants or grid extensions will become more and more complicated. Furthermore, this type of capital-intensive structure could hardly be implemented for comprehensive electrification in developing countries. There are several ongoing approaches to develop novel grid structures which enable both the optimization of existing grids in industrialized countries and ecologically acceptable electrification in threshold countries. These approaches aim to help in the transition to a decentralized structure of small- and medium-sized generators. In these concepts new devices like cogenerations sets, fuel cells, and renewable energy systems like photovoltaics, wind, micro hydro, biomass, etc. can play a key role (Fig. 2.43).

However, due to specific operational constraints all these components, new devices must be incorporated in such a structure to achieve optimized use of these new technologies. Short-term energy storage consisting of locally installed batteries or flywheels, power quality management (also on the low-voltage side to reduce voltage breakdowns and reduce harmonic distortion), and intelligent communication systems between individual components and a central energy management unit have to be applied.

Future Developments in Grid Connected Photovoltaic Systems Grid-connected photovoltaic systems are being introduced in large numbers in several national programs, like the 70 000 Roofs-Program in Japan or the

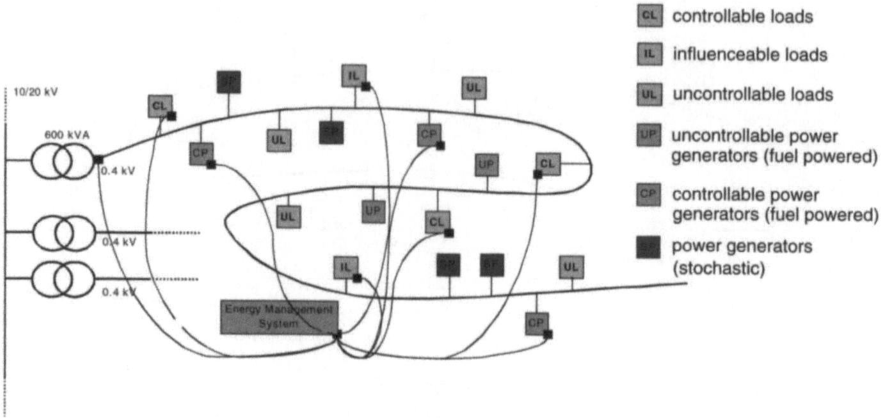

Fig. 2.43. Future decentralized grid structure with superimposed communication network

100 000 Roofs-Program in Germany. In other countries, like the Netherlands, Spain, or the United States, similar markets are under development. Besides the rooftops, more and more large central power stations are being erected by utilities or industries on facades, roofs, or in the flat field. Besides the pure aspect of electricity production, grid-connected photovoltaic systems have an enormous affect in public relations, for example for utilities and enterprizes. Therefore, quality control, monitoring, and dissemination of the results gained with these installations are key factors for the success of these installations and their further dissemination. To judge by the development of grid-connected PV systems in the last few years in Europe, photovoltaics can be expected to be one of the driving forces towards the reduction of CO_2 emission in industrialised countries.

As a result of the liberalization of the European electricity market and electricity distribution, grids are subject to significant technical and organizational changes. The process has just started and will continue for years. Major results of the liberalization are the organizational separation of power generation, transmission, and distribution companies, and supply companies and the open market for all participants, resulting in fierce market competition and price decline. This again will lead to an increasing number of distributed smaller and medium-sized power generators, an increasing replacement of traditional means of power generation (like nuclear power) by a mix of different technologies (like combined heat and power units, renewable energies), and partly increasing demands of the customers on the electricity quality and availability (e.g. sensitive electronics, computer systems, telecommunication).

To cope with these new market conditions, it is necessary to give utilities and grid operators cost-efficient planning tools to make them capable of including distributed power generators (like renewables) in a way that will contribute to power demand and save generation capacity for the utility. In addition, the new management schemes will require flexible adjustment to the needs of specific customers with respect to power quantity and quality, e.g. premium power, individual treatment of different customers according to their requirements and willingness to pay. For higher quality and reliability, higher prices will have to be charged. The decentralized character of renewable energies may give economic advantages when calculating the avoided costs, for example for grid extension or grid upgrading.

2.3.5 Fluctuations of Photovoltaic Power Production and Storage Requirements

In the time domain, PV power production is characterized by (i) stochastic short-term fluctuations, generally due to moving clouds; (ii) medium-term trends caused by variations of general weather situations; and (iii) deterministic diurnal and seasonal variations.

Future electricity supply systems that rely on electricity from fluctuating sources (e. g. PV and wind-generated electricity) have to cope with these stochastic and deterministic variations, both with respect to the grid structure and the complementary electricity generation.

Solar short-term fluctuations are local effects; the statistical correlation of these fluctuations decreases strongly with the spatial distance of the PV stations. Thus the overall fluctuations in large-area grids, which have a large number of PV electricity inputs, will be leveled out considerably. Qualitatively this may be seen in Fig. 2.44, which shows the lumped energy production of 100 PV generators dispersed approximately evenly over Germany.

This statistical smoothing [Wiemken 2001] will result in a reduction of overall fluctuations to a level that can be compensated via the control of complementary electricity generators (e. g. fossil, hydro, biomass). A prerequisite for this is the existence of an appropriate grid structure: large area (preferably continental) grids with a pronounced bidirectional power transport capability. The latter requirement may be reduced by the introduction of dispersed storage facilities, which have already been discussed in distributed power generation schemes.

Using these measures, short-term fluctuations (up to modest fractions of an hour) will not pose a fundamental technical problem for large-scale application of PV electricity production, provided a sufficient fraction of electricity (see below) is produced by means of power plants that can be controlled over characteristic times of approximately half an hour.

In prinziple, diurnal and seasonal variations can also be handled by large bidirectional grids. In this case, high-efficiency global links for electric power – utilising superconducting or high-voltage DC energy transport, for example – would be necessary. Since it is not clear whether such systems can be installed globally in due time, it is essential to assess the fraction of fluctuating electricity from renewable energy sources that can be handled by large

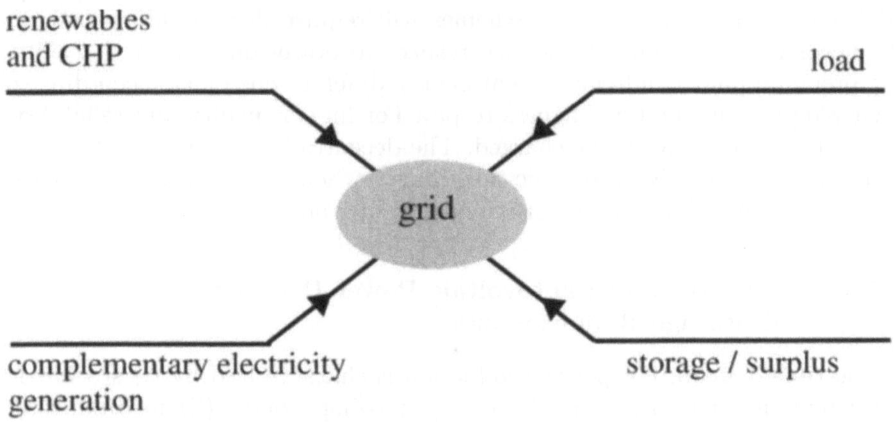

Fig. 2.44. Structure of model calculations for the assessment of the necessary complementary electricity generation capacity and storage requirements for grids characterized by a high penetration with electricity from fluctuating renewable energy sources. For simplicity it is assumed that the grid exhibits ideal bidirectional behavior. Thus the energy balance can be calculated at a single point. The production of electricity via combined heat and power (CHP) generation units (not mentioned in the text) is calculated together with the input from renewables since the operation strategy of CHP is strongly influenced by renewable heat energy production. Input from renewable energy sources comes from PV (local and import), wind (including offshore), solar thermal power plants (import) and hydro (including storage plants). For further details see [Langniß 1998]

(but not global) electricity supply systems. This problem can be analyzed by means of simulation calculations.

The gross structure of such a model calculation for Germany is shown in Fig. 2.45 [Langniß 1998]. For each time step of the simulation the difference between load and generation from renewables yields the power that has to be produced by means of complementary power plants (e.g. fossil-fueled installations). In cases where the electricity production from renewables surpasses the demand of the loads, surplus energy is produced; this may in principle be fed to storage devices.

Such simulations are based on essential stipulations of parameters and time patterns. For the example discussed below these are: the mix of renewable energy converters, large-scale time- and space-resolved meteorological data, and load patterns. The example given here assumes amongst others the following annual electricity production capacities:

PV Germany	80 TWh
PV import from southern countries	50 TWh
Wind Germany	80 TWh
Total electricity production	599 TWh

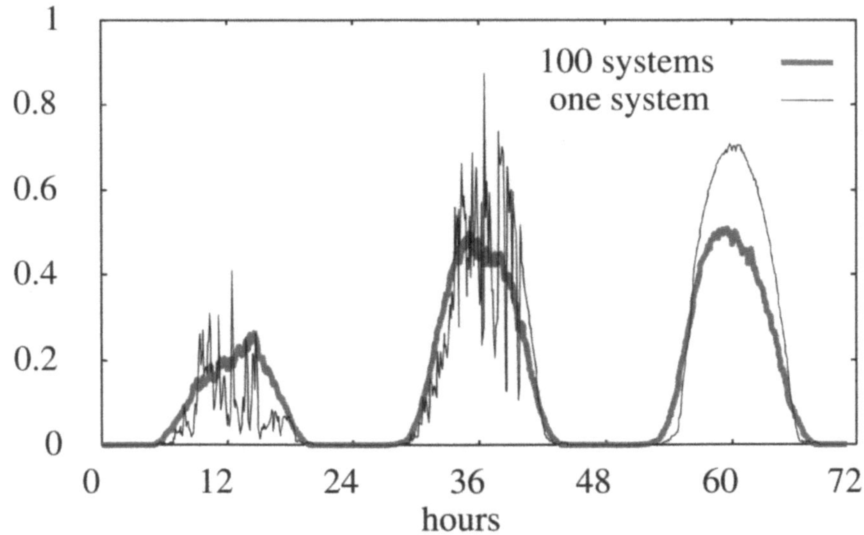

Fig. 2.45. Power output of one and 100 dispersed PV systems [Wiemken 2001]. Experimental data (three days in June 1999) from the 1 000 Roofs Program, Germany. The output power is normalized to the installed nameplate power of the PV systems. Time resolution: 5 minutes

Thus in this futuristic scenario it is assumed that 35 % of the annual electric energy is produced via locally highly fluctuating renewable energies. Meteorological data with sufficient spatial resolution are available at hourly resolution only. Thus it is implicitly assumed that fluctuations below this time span are statistically leveled out or smoothed by the application of dispersed storage devices. As a first step it is assumed that the time structure of the load is the same as today.

The consequences of such a scenario are displayed in Fig. 2.46. The time pattern of the complementary electricity production shows almost no base load behavior. These power plants are utilized to level out the remaining variations of the power balance in the supply system. Twenty percent of the annual electricity is produced by these units. On the other hand, strong surplus generation occurs. It adds up to 33 TWh, which is equivalent to 10 % of the yearly electricity production from all renewables (see caption to Fig. 2.44). Without large-scale storage facilities, this surplus energy would constitute a loss.

The simulation result displayed in Fig. 2.46 is based on the assumption that the load pattern is not adapted to the time pattern of renewable energy production. On the other hand, for such scenarios it would be consequent to use methods of load management to shift the load peak to noon, and thus have better correlation between energy production and consumption. This may be done, for example, via the following measures: shifting a fraction of

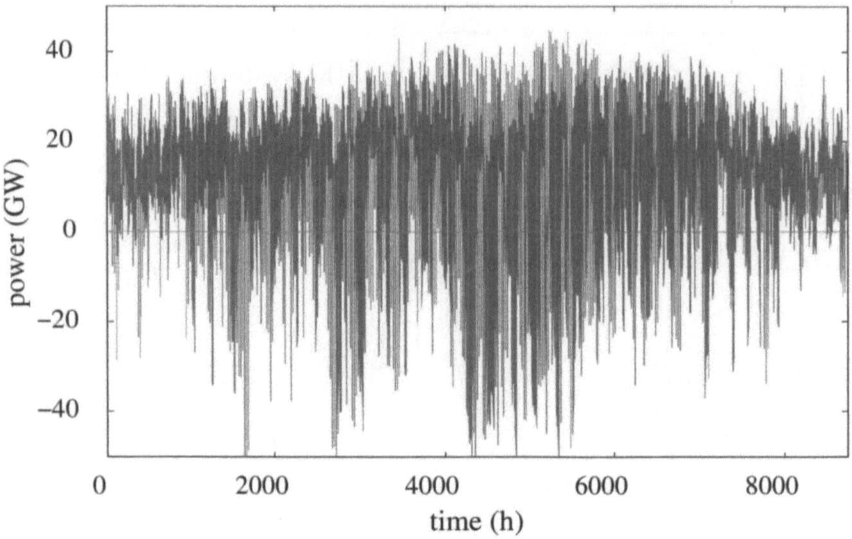

Fig. 2.46. Balance (see Fig. 2.44) of electricity power input from renewables (including CHP) and the load for the scenario described in the text. Positive values signify necessary power input from complementary controllable plants, negative values indicate surplus power or power fed to large storage systems respectively. Total installed generating capacity (renewables and fossil-fueled) is 251 GW

the electricity demand of households (charging of cooling devices, operation of washing machines, etc.), charging of electric vehicles at noon, and adapting the time profile of CHPs and pumped storage power stations (for quantitative details see [Langniß 1998]).

This results in a power balance time pattern as shown in Fig. 2.47. Both the power production from complementary plants and the surplus generation (or storage requirements) are reduced considerably, the latter now being approximately 3 % of the total renewable electricity production.

From these simulation calculations, which do not constitute a systematic analysis of the problem, it may be concluded that it is technically feasible to feed at least 35 % of highly fluctuating electricity from PV and wind sources into the German electricity grid. This figure is based on the assumption of a strong, bidirectional national electricity grid and stringent load management measures. The figure may become even higher if continental or intercontinental grids are taken into considerations. The introduction of more elaborate load management strategies utilizing e.g. distributed electronic intelligence in appliances would work in the same direction. From a strategic point of view these findings suggest that in parallel with the installation of PV electricity generation capacity, the grid structure, the system of complementary electricity production, and the load management technologies must be adapted to the new energy supply scheme.

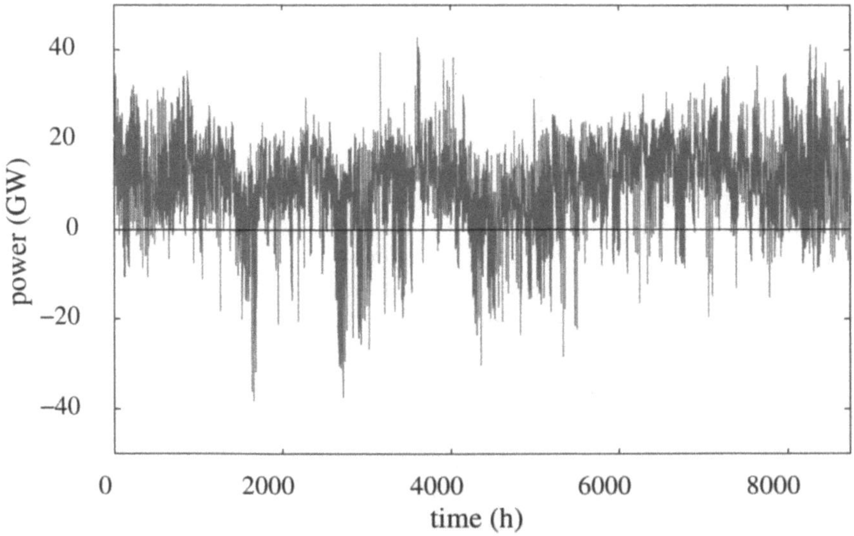

Fig. 2.47. Power balance as shown in Fig. 2.46 but based on a load pattern that has been adapted in its time behavior to the characteristics of solar electricity production (see text). Total installed generation capacity remains 251 GW

References

[Amann 1996] Amann, J. Gieshoff, J. Kuhmann, R. Kügele, W. Roth, W. Schulz, A. Steinhüser (1996) Produkte mit integrierter photovoltaischer Energieversorgung, BMBF-Statusseminar Photovoltaik, Bad Breisig, 23.–25.4.1996, S. 43–1 bis 43–20

[Asia 1996] Asia Alternative Energy Unit ASTAE (1996) Best Practices for Household Electrification Programs: Lessons from Experiences in Selected Countries, World Bank Technical Paper, Number 324 (1996)

[Aul 1995] H. A. Aulich, Proc. 13th EUPVSEC, Nice, 1995, 1441–1444

[Ber 1999] R. B. Bergmann, Appl. Phys. A69 (1999) 187–194

[Bet 1999] A. W. Bett, F. Dimroth, G. Stollwerck, O. V. Sulima, Appl. Phys. A69 (1999) 119–129

[Bon 1992] B. Bonnet, Int. J. Solar Energy, 12 (1992) 1–14

[Bopp 1997] G. Bopp, H. Gabler, K. Kiefer, K. Preiser and E. Wiemken (1997) Hybrid Photovoltaic-Diesel-Battery Systems for Remote Energy Supply, North Sun '97, Espoo, Finland (1997)

[Bopp 1995] Bopp, G. Bächler, M., Gabler, H., Höhe, W., Mittermeier, J., Sauer, D.U., Sprau, P., Willer, B., Wollny, M. (1995) A systematic effort to define evaluation and performance parameters and criteria for lead-acid batteries in PV systems, 13th European Photovoltaic Solar Energy Conference, Nice, France, 1995

[Butera 1989] Butera, F. (1989) Renewable Energy Sources in Developing Countries: Success and Failures in Technology Transfer and Diffusion, Progretto Finalizzato Energetica, Rome, 1989

[Erge 2000] Th. Erge, K. Kiefer, V.U. Hoffmann (2000) The German experience with grid-connected PV systems, Solar Energy Journal, 2000

[Erge 1998] Th. Erge, G. Heilscher, B. Genennig, U. Rindelhardt, G. Blässer, H. Rieß, B. Decker, H. Decker et al. (1998) The German 1 000-Roofs-PV-Program: a Résumé of the 5 Years Pioneer Project for Small Grid-Connected PV Systems, 2nd World Conference and Exhibition on Solar Energy Conversion, Vienna, Austria, 6.–10.7.1998, S. 2649–2651

[Foley 1990] Foley, G. (1990) Electricity for Rural People, Panos Publications, 9 White Lion Street, London N1 9PD, UK (1990)

[Foley 1995] Foley, G. (1995) Photovoltaic Applications in Rural Areas of the Developing World, World Bank Technical Paper, Number 304 (1995)

[Gabler 1995] Gabler, H., Wiemken, E. (1995) How much Energy will a Specific PV System Produce? Proc. 13th European Photovoltaic Solar Energy Conference, Nice, France, 1995

[Gabler 1998] Gabler, H., Preiser, K. (1998) Technical Quality Control for Solar Rural Electrification Institutional Cooperation for Solar Energy Activities in Mekong Riparian Countries, Hanoi, Vietnam, 1998

[Goe 1998] A. Goetzberger, J. Knobloch and B. Voss. Crystalline Silicon Solar Cells, J. Wiley & Sons, 1998

[Glu 2000] S. W. Glunz, B. Köster, T. Leimenstoll, S. Rein, E. Schäffer, J. Knobloch, T. Abe, Prog. Photovolt. 8 (2000) 237–240

[Gra 2000] M. Graetzel, Prog. Photovolt. Res. Appl. 8 (2000) 171–185

[Gre 1986] M. A. Green, Solar Cells, UNSW, 1986

[Guh 1999] S. Guha, J. Yang, IEEE TED 46 (1999) 2080–2085

[Heydenreich 1996] Heydenreich, W, Wiemken, E., Gabler, H., Kiefer, K. (1996) Correlation between photovoltaic energy production and household electricity consumption evaluation of data from the German Thousand Roofs Program, Proceedings EuroSun 96, Freiburg, 16.–19.9.1996

[Hille 2000] Hille, G., Preiser, K. (2000) Innovative Financing Scheme: Joint Ownership of Photovoltaics, ALTENER-Conference 2000, Toulouse, France

[Kleinkauf 1994] Kleinkauf, W., Sachau, J. (1994) Components for Modular Expandable and Adaptable PV Systems Proc. 12th European Photovoltaic Solar Energy Conference, Amsterdam, Netherlands, 1994

[Kleinkauf 1996] W. Kleinkauf and F. Raptis (1996) Elektrifizierung mit Hybridanlagen – Einsatz erneuerbarer Energien zur dezentralen, netzkompatiblen Stromversorgung, Forschungsverbund Sonnenenergie, Annual Conference (1996)

[Kleinkauf 1998] W. Kleinkauf, J. Bishop, O. Haas, F. Raptis, J. Sachau, P. Zacharias; (1998) Decentralized Grid-Compatible PV Power Supply – Strategy For Electrification and Integration Concepts With Modular Technology, 2nd World Conference on PV Solar Energy Conversion, Vienna, July 1998

[Langniß 1998] Langniß, O., Nitsch, J., Luther, J., Wiemken, E. (1998) "Strategien für eine nachhaltige Energieversorgung", Workshop des Forschungsverbundes Sonnenenergie, Freiburg, 1997, ISSN 0949–1082

[Linden 1995] Linden, D. (1995) Handbook of Batteries, 2. Dition, McGraw-Hill, 1995

[McNelis 1995] McNelis, B., Barlow, R. (1995) Photovoltaic Water Pumping, Hille, G., Roth W., Schmidt H., PV Systems, Seminar Handbook, Fraunhofer ISE, Freiburg, 1995

[Mey 2000] P. V. Meyers, S. R. Albright, Prog. Photovolt. Res. Appl. 8 (2000) 161–169

[Munro 1995] Munro, D. K., Hacker, R. J., Blaesser, G., Riesch, G. (1995) The Long-Term Performance of PV Systems and Components, Proc. Solar World Congress, Harare, Zimbabwe, 1995

[Njaimi 1992] Njaimi, M; Ounalli, A.; Ullerich, U.; (1992) Solar Home Systems: an Alternative to Rural Electrification Programs via the Electric Grid, 11th E.C. Photovoltaic Solar Energy Conference, Montreux, Switzerland, 1992

[Perodi 1995] Parodi, O.; Preiser, K.; Schweizer, P (1995) Balde de Leyes: The integrated way to electrical light, 13. EU PV solar energy conference; Nice, France, 1995

[Perodi 1997] Parodi, O, Preikschat, M. Preiser, K. (1997) PV contra Coli-Bacteria – Suitability of UV Water Purification Devices for PV Systems, 14. EU PV Solar Energy Conference; Barcelona, Spain, 1997

[Perodi 1998] Parodi, O.; Preiser, K.; Schweizer-Ries, P., Wendl. M. (1998) When Night Falls on Balde de Leyes – The Success Story of an Integrated Approach in PV Rural Electrification, 2nd World Conference and Exhibition on Photovoltaic Solar Energy Conversion, Vienna, 1998

[Pfanner 1995] N. Pfanner, W. Roth, A. Steinhüser (1995) Application of Photovoltaics in the Stationary Sector of Road Traffic, 12th European PV Solar Energy Conference, Amsterdam, 11.–15.4.1994, S. 206

[Preiser 1995] Preiser, K., Parodi, O., J. Kuhmann (1995) Quality Issues for Solar Home Systems, 13. EU PV Solar Energy Conference, Nice, France, 1995

[Roth 1996] W. Roth, J. Schilz (1996) Thermoelectric Generators as Auxiliary Electricity Generators, Research Association for Solar Energy, Proceedings Annual Conference 1996, Kassel, Germany 1996

[Roth 1999-1] W. Roth, H. Lemmel,(1999) Solarversorgte Kleingeräte – 1. Teil, "Elektronik" Fachzeitschrift für industrielle Anwender und Entwickler (Nr. 18, September 1999), S. 84–88 (1: Journalist für Technik und Wissenschaft)

[Roth 1999-2] W. Roth, H. Lemmel (1999) Solarversorgte Kleingeräte – 2. Teil, "Elektronik" Fachzeitschrift für industrielle Anwender und Entwickler (Nr. 20, September 1999), S. 40–43 (1: Journalist für Technik und Wissenschaft)

[Sauer 1997] Sauer, D.U., Bächler, M., Bopp, G., Höhe, W., Mittermeier, J., Sprau, P., Willer, B., Wollny, M. (1997) Analysis of the performance parameters of lead/acid batteries in photovoltaic systems, J. Power Sources, 64 (1997) 197–201

[Schock 2000] H. W. Schock, R. Noufi, Prog. Photovolt. Res. Appl. 8 (2000) 151–160

[Schubnell 1997] Schubnell, M., Gabler, H., Broman, L. (1997) Overview of European Activities in Thermophotovoltaics, 3rd NREL Conference on Thermophotovoltiac Generation of Electricity, Colorado Springs, USA, 1997

[Schweizer 1995] Schweizer, P. (1995) Psychologische Faktoren bei der Nutzung regenerativer Energien. Eine Studie zum Einsatz von Solartechnik im zentralen Himalaya, PhD thesis, University of Heidelberg, 1995, ISBN 3-931660-51-6

[Steinhülser 2000] Steinhüser, W. Roth, W. Schulz, H. Schmidt (2000) Photovoltaic Power Supply for Telecommunication Network Components in Remote Areas, Proceedings Telescon 2000 – The Third International Telecommunications Energy Special Conference, 7.–10.5.2000, Dresden, S. 221–225

[Strat 2000] Strategies Unlimited (2000) Report PM 48, San Antonio Circle, California, USA 2000

[Swan 2000] R. M. Swanson, Prog. Photovolt. Res. Appl. 8 (2000) 93–111
[The World Bank 1996] The World Bank (1996) Rural Energy and Development Improving Energy Supplies for Two Billion People, The World Bank, Washington, 1996, ISBN 0-8213-3806-4
[UN 1998] UNDP United Nations Development Program (1998) Human Development Report 1998
[Van 1986] R. J. Van Overstraeten, R. P. Mertens, Physics, Technology and Use of Photovoltaics, Adam Hilger, (1986) 131–142
[WHO 1993] WHO- World Health Organisation (1993) Guidelines for Drinking Water Quality, Volume I, Recommendations, Geneva, Switzerland, 1993
[Wiemken 2001] E. Wiemken, H. G. Beyer, W. Heydenreich and K. Kiefer "Power characteristics of PV ensembles: experiences from the combined power production of 100 grid connected PV systems distributed over the area of Germany," Solar Energy, 70, 6, 2001
[Wiengmoon 1997] Wiengmoon, B. Wattanapong R., Parodi, O., Wiemken, E. (1997) PV-Battery Charging in the Huay Dua Village, Thailand, South-East Asia PV Conference, Phuket, Thailand, 1997
[Wilk 1992] H. Wilk, (1992) "Utility Connected Photovoltaic Systems, Power Conditioner and Utility Interface", IEA – Solar Heating and Cooling Program, Task 16, "Photovoltaics in Buildings," Design Handbook, October 1992

3 Industrial Technology and Cost Reduction Assessment

T. Schott

3.1 Photovoltaics as a Technology and Industrial Development

To evaluate the potential of cost reduction and its possible realization for industrial photovoltaic production, the following questions have to be answered:

- What is the possible speed of technological progress concerning production cost and what conditions are required?
- What is the possible rate of technology transfer to manufacturing?
- What is the speed of market development in relation to investment cycles in fabrication?

and after all, if there are useful answers to all these questions,

- How do the answers finally fit into a strategic scheme?
- What are the social and political boundary conditions for this process?
- What are the cost (public od private) to accelerate this process?

These questions are not new and historically have been put several times. There have been given optimistic as well as "cloudy" forecasts for Photovoltaics [DeMeo 1984, Norman 1984] reflecting the fact that the specific cost of PV has dramatically fallen by the 1980s, but still suffered from a large gap to full commercial competitiveness (apart from a growing number of special off-grid applications). These early commentators clearly demanded political action for a further continuous long-term-support of PV development.

To approach these questions again, in this chapter the situation of photovoltaics at the industrial scale is to be analyzed from a present status and conclusions have to be drawn for possible strategies how to realize the anticipated cost reduction potential.

For practical reasons, the analysis considers two time horizons, a first one up to 2010 and a second at 2020. Beyond 2020 the perspectives at present are rather speculative.

Any outlook beyond 2020, however, demands more vision than analysis, which is of course also necessary within the political scope. It is assumed that policy has degrees of freedom to decide upon. The question is much more about the present possibilities and sensibilities to steer the system, wherever political reason is guided to. And in this context, it is of highest importance, what actions are required to be taken now.

The situation can be compared with the control of a big ship. The bigger the ship the earlier any control movement has to be activated to keep the course or to set a new course. This means that the present analysis does not stop at 2020, but is only devoted to draw attention to the effectiveness of measures available today compared to the vision that photovoltaics will play an important role in the long run which means in the second half of this century. This is not a new situation as can be seen from the nuclear case, where original visions of "atoms for peace" went even for centuries to come. However, as can also be experienced from the nuclear case, long visions also bear the risk of a wrong course. Therefore, we should be careful not to overstrain the potential. The next twenty years will show if PV is a real vision, not only technical but also economical.

For these two horizons, the quality of conclusion naturally will differ. Up to 2010, only the technologies already at the market or just entering have a chance. For them, rather conservative estimates can be made.

On the other hand, already now industry is looking beyond 2010 judging future R&D engagement. Therefore, beyond 2010, also opportunities now visible and emerging from the laboratory level have to be included. Of course, the data basis for these newcomers as well as the competing prospects for the already participating runners is much less certain.

The time horizons correspond to the medium and long-term goals for photovoltaics.

Coming from economy, a reduction by a factor of 10 in specific cost (whatever currency per W_p) may be set as a reasonable goal to make PV a global energy option and the following steps could be set:

Cost reduction: 1/2 by 2010 and another factor 1/5 by 2020

If we stick to the historical learning curve (see also the attached chapter excursion on the practice of learning curves) during the next 10 years and we expect a factor 1/2 in costs, then the production must grow at an average yearly rate of 26 % p.a. (to arrive at a tenfold market volume in 2010 compared to 2000).

Compared to the average growth rate during the last decade, which resulted in about 15–20 % average annual growth rate, the acceleration of the development process seems to be necessary. It must be stated that the market growth recently has been increased to values around 30 % p.a. It is, however, not clear whether such a growth can be continued for the next decade and whether it follows the historical learning curve experience.

This is exactly a position which may be quoted from a number of recent sources of discussion. For instance in [NCPV 1999] the strategic goal is stated by the US-PV industry with the following vision:

"To realize a thriving US-based solar-electric power industry, which provides competitive and environmentally friendly energy products and services that meet the needs and desires of the domestic electric-energy consumer" leading to four strategic action tasks:

1. Maintain the US industries worldwide technological leadership
2. Achieve economic competitiveness with conventional technologies
3. Maintain a sustained market and PV production growth
4. Make the PV industry profitable and attractive to investors

From non-US point of view this could be considered somewhat too ambitious (if not discriminating against the rest of the world). However, this is exactly the situation where the PV industry now stands. There are being formed strategic goals and alliances as well as strong efforts taken to develop the markets and to conquer them early. The competition has been starting in all industrially leading countries. So one important boundary condition of industrialization of PV beyond the technological process is a strong competition of national interests (perfectly stated above by the US industry).

If the national interest is taken as a similar commitment as holding for other commercial industrial areas (especially when export oriented as for airplanes, electronics, etc.) the argument leads to the following statement:

> Industrialization of PV must be considered a strategic goal in a global competition which cannot be left simply to free-market forces. The "free market" does not guarantee entry into the game as a main player in later stages.

As a consequence, the PV strategy has to be discussed at a level of industrial policy apart from its other benefits as e.g. for the environment or for social demands of the developing countries. The latter, however, are important in the way, that PV (and in general renewable technologies) feed and justify the hope, that these human demands can be covered together with and not in contrast to economic advantage.

The US "solar million roof" initiative reflects this ambitious approach by setting goals in production capacities and cost development within this decade [DOE 2000] considering also PV a job creation machine (see Table 3.1). Such goals can be taken as typical for the next development era of PV until 2010 exhibiting the technical and economical potentials and expectations. However, the US program is still questionable at the governmental level and has not yet reached a comparable status like the German 100 000 roof program or the Japanese programs. Also a number of other countries (e.g. Netherlands) are considering new market introduction and industrialization programs.

It is now important to judge the possible forecast for the necessary market growth. A number of facts and foreseeable boundary conditions seem to provide such growth over the next ten years. Other facts are not automatically

Table 3.1. US Million Solar Roofs Goals and Benchmarks for Industrialization

	1999	2000	2005	2010
Participating Partners	25	50	200	325
Solar Buildings and Residences[a]	23.5	51	376	1 014
System Size (kW)	1	2	3	4
Annual Capacity (MW)	15	55	270	610
Total Installed Capacity (MW)	25	80	820	3 025
Installed Cost ($/W)	4.9	4.3	2.9	2.0
Energy Cost (c/kWh)	16.9	14.8	10.6	7.7
Total Annual CO_2 Savings[b]	39	111	1 037	3 510
Jobs Created[a]	3.8	11	40	71.5

[a] in thousand
[b] thousands of tons

in favor of such growth. Both shall be discussed in this chapter in relation to the anticipated industrial development area.

The PV history is now looking back to a quarter of the past century with commercial market development. This together with the fact that its learning behavior during that period fits quite well into the general scheme of technological evolution with usual parameters justifies a reasonable forecast for at least one decade if not for two.

Also, if one is convinced (as the majority of industries in the PV area are) that the battle between the emerging new technologies, the thin-film technologies and the "workhorse" crystalline silicon wafer technology will dominate the market after 2005, then a sound basis of technological information for the learning process is already guaranteed until 2020.

Trying to look beyond 2020 means that we have to take into account the progress now just coming into play at the laboratory level. The results, however, may be visible and perhaps may be verified on a pilot scale after 2010 but will not come into industrial play before then. Instead according to the necessary learning steps to become a real industrial option, they will need another decade to be involved at the large scale of production with substantial market shares after 2020. This may be another generation of thin-film technologies, like organic PV cells, flexible cells, low temperature thin-film deposition processes, etc.

It is clear that the whole development process over more than two decades still requires great efforts in research and technology to reach the (industrial) goal of cost reduction for PV. It is by far not automatically the case and also not guaranteed. Research results can be imagined but not guaranteed by definition, and require a good foundation of their own, and especially long-term continuity. Therefore, whatever the strategy on the industrial side will be, it requires a balanced complementary strategy on the side of academic research and technology transfer.

Taking into account the dramatic globalization and concentration speed of high-tech industries, this leads to a hard competition not only between global companies, but also increasingly between the industrial nations (and also between regional industrial areas).

If there is a promising new market with new key technology, generally a technological and industrial race is the consequence. To become a winner in such a technological race, it is inevitable to be an early competitor with a sufficient "staff and support" to get the chance. In the latter case, it is especially the boundary conditions of intrinsic industrial abilities given by the regional conditions where the industry shall be allocated. This defines the attraction of a given industrial region with respect to new technological approaches (the Silicon Valley effect).

There is now growing conviction that photovoltaics provide a quite general key technology of the 21st century worth the effort of thorough industrialization. Therefore, to open a chance to become a winner of the game, appropriate boundary conditions for this automatic competition and acceleration of the development process are requested. It would not be a reasonable strategy to wait and see how the photovoltaic market will develop and to try to enter the market after it has developed.

3.1.1 PV Production as an Industry of Its Own Kind

For the first time, after almost 50 years of photovoltaic history, the PV industry with a global turnover of more than $1 billion US has reached a stage where the level of production is of the order of global investment strategies. This has the consequence that investments are no longer merely guided by criteria of technologists but also of professional economists.

On the other hand, photovoltaics are still an emerging technology with enormous potentials at one hand and with efforts of development on the other side. Therefore, for professional investors, the relation of future market expectations to the speed of technological progress is decisive.

The struggle between technological progress on one hand and the economical analysis and the market conditions on the other hand often lead to misunderstandings concerning the present and future status of photovoltaic technology. This was the more the case as photovoltaics already represent a fully economical feasible opportunity for a number of niche markets, where special and selective technical advantages are of primary importance.

Despite all the complexities of the inherent technology development, photovoltaics have shown remarkable steady market development. During the last 10 years, the installed PV generator power has grown with a rate of about 15 % p.a. and with cost reductions (taking PV modules as significant) of about 7 % p.a. This, however, means that the turnover of the PV industry has grown at a rate of about 8 % p.a., which is not bad but also no real highlight from an economist's point of view.

There are lessons learned from the case of the electronics industry.

For several reasons, it is tempting to compare photovoltaics with learning processes in similar cases and especially, of course, with its root technology, the development of silicon-based electronics.

The following similarities can be cited:

- PV is a new technology and a fundamental innovation making use of the same solid-state physics as applied in the case of electronics and of similar production processes
- A solar cell is a "smart" energy conversion device providing often completely new solutions to a number of technical problems
- Solar cells are modular and applicable in an extremely wide range of scale
- PV is fascinating and also deeply convincing for the end consumers, producing direct utility electricity from sunlight by obviously simple means and without any moving parts
- There were also strong incentives from early economic applications, the most important in space flight in both cases
- It is also interesting to note that the industries involved classically emerged from companies dealing with electronics and information technologies. The following properties make PV an economically interesting technology in a similar way.
- It is a rapidly growing new branch of industrial production
- It requires highly sophisticated machinery at a level similar to electronics
- Technology allows large mass production with high cost-reduction potential

However, *fundamental differences between photovoltaics and electronics* industries also have to be taken into account:

- In competition with conventional power supply (i.e. in the sense of PV as a global solution), in principle, a PV generator does not provide more than simply kWh of electricity produced, whereas electronic computers opened the road to completely new applications and market opportunities not comparable to any technical quality or service before. Energy is not a new objective in industrialized societies.
- Electricity produced from photovoltaic generators, if competing with bulk electricity within the public power supply in large grids, shows even disadvantages concerning the necessity to be combined with storage or at least conventional power supply with cheap and easy storable primary energy input.
- The progress in electronics is mainly due to higher speed and miniaturization of transistors together with integrated production of highly complex microstructures on small chips of silicon crystal. In contrast to this, solar

cell devices are fundamentally bound to receive solar radiation on a given aperture area (even highest physical efficiencies provide no real new order of magnitude). So it remains a geometrically large-scale technology and there is no escape from this area-related physical condition.

Essential new qualities of PV are its environmental friendliness, its off-grid use at arbitrary scale, and its unique applicability to real needs in developing countries. These qualities, however, are general advantages, not so easy to be transformed into economic advantages.

A remark concerning PV concentrator technologies is appropriate. In principle, concentrators may shift the scale problem from the cell itself to another part of the total cost, namely the structural elements around the PV cell. So, the cell area may be designed smaller with increasing concentration of direct solar radiation. The cost partitioning is then deferred more to the components necessary to concentrate and to the (inevitable) tracking devices. Nevertheless, the same basic questions still have to be answered again for these systems. In principle, PV concentrators allow more sophisticated cell design and thereby also access to higher efficiencies.

In conclusion of this short comparative analysis, if there is production cost to be reduced, it means that *for the PV economy the added value per some squaremeter of structure has to be reduced and the efficiency of power output has to be maximized.*

Now the big difference to electronics becomes clear. For electronics, it is possible to increase the added value per unit area of structural material by increasing complexity and function per unit surface area of a silicon wafer and indeed has been increased drastically by orders of magnitude during the decades of electronic development leading to its historical dramatic price fall.

If we take these differences into account, the future PV industry inevitably has to free itself of its electronic past and to end up as an industry of its own, perhaps far more similar to the glass industry. It will thereby become bound to final basic material requirements different from any other power technology existing today.

The potential of PV in the long run will be connected to other questions like "what is the lower physical limit of structural mass per unit area of solar generator aperture or what will be the ultimate technically feasible efficiency of solar cells and, of course, what will be the ultimate manufacturing processes as simple and cheap as possible?"

Concerning these latter questions about progress, a warning is recommended [Schmidt-Tiedemann 1997]:

In the history of new technologies, often the early estimations on physical limits have missed their goal as well in pessimism as in optimism. The retrospective reason for that uncertainty is mostly found in a typical neglect of the (practically unforeseeable) interwoven general progress in quite different areas of technologies, which have to be merged to result in a complex technological approach. Even the big success of electronic computers is an example, which has been completely underestimated in the beginning. Really no one had anticipated the enormous achievements in miniaturization together with the progress in basic solid-state physics and engineering.

So the ultimate questions and analogies should be dealt with rather carefully to avoid premature conclusions. Instead it appears more suitable to look for appropriate boundary conditions opening the door to technology and market development.

To draw attention to these questions nevertheless is quite helpful from the start, bringing to light basic fundamental restraints and key factors to be recognized as early as possible by the engineer, researcher, and economist.

Already during the relatively young history of industrial photovoltaics, some basic misinterpretations of this kind took place. A good example is the story about the energy payback time (cf. Chap. 4) of photovoltaic solar cells and systems. In the beginning of the first commercial productions, solar cells showed a rather bad balance of input to output energy with payback times of many years, justifying a sceptical questioning, if there could be achieved a reasonable balance at all.

Of course, the positive counter argument is derived from the fact, that in these early pilot productions the quality of the product had absolute highest priority and that the cost for input energy taken from (relatively cheap!) conventional supply could be ignored for pilot productions.

In this example, the (absolutely justified) question on the total life cycle energy balance has to be inverted as a task for engineers and investors to take care of the input energy demand for a future production technology at large scale. If photovoltaics will become as cheap as anticipated, then the production has to "learn" how to reduce energy cost and automatically will lead to positive balance. There is no fundamental physical limit to this improvement. As a consequence, it is not justified to draw a negative answer from the present status of technology.

However, long before that end is reached, industries under competitive pressure have to follow realistic paths permanently securing a reasonable balance between technical and economical progress. From this, as in all technological developments, the success of industry depends on finding appropriate steps of innovation together with realistic estimates of the market development.

3.1.2 Industrial Goals

Concerning the above problems, the real technical and economical goals the PV industry is facing are quite different from the tasks of electronic development concerning markets and technical means. As long as a PV module can be produced with standard specification independent from the system, we can define a separate PV module industry.

Specific goals for a future PV module industry are then:

- *Maximize* the annual electricity production from a given aperture area to be used with a solar generator at a given site and under given meteorological conditions, which in general means increase of efficiency of the solar cells (modules, strings, etc.)
- *Minimize* the specific production cost per unit of module area output, which leads to the increase of production capacities, yield, processing speed, standardization, etc.
- *Establish mass production* with module sizes up to the range of sqare meters
- Evaluate new pilot technologies beyond the present typical module design (e.g. flexible substrates, etc.)

There are additionally further and more subtle goals of adaptation to specific market segments of applications. This is related to all cases in which PV is applied to special applications by integrating and adapting the photovoltaic active area into surfaces originally providing other services, as for instance for facades, roofs, shadings, mobile equipment, etc. In the latter case, other relations and requirements as density, integrability into surface geometry, optical performance, etc. may dominate the economy.

If such specific market segments are big enough to justifiy a unique PV application, a complete system approach is required, including power conversion, storage etc. The value for the customer to be maximized in this case is no longer simply electricity, but becomes a mixed and unique demand to be described by inherent specifications. Such specifications then go beyond the grid-integrated PV systems and in most cases include important *non-energetic* parameters for the system performance (e.g. aesthetics in architectural design).

Despite the fact that in many PV applications the output performance of PV modules can be optimized independent from the connected system components, this does not hold everywhere. For instance, in the case of facade-integrated PV, the influence of photovoltaic conversion efficiency in relation to the added cost of additional photovoltaic power production is quite different compared to a separately installed generator.

Nevertheless, the primary solar cell efficiency remains a fundamental parameter of economic scale. For comparable PV technologies in general a higher efficiency, if achievable at equal cost, also tends to be more economic for the area-related cost share. This has led to a permanent racing towards higher cell efficiencies in basic research and also to steadily improved standard efficiencies in manufacturing (compare for the history of cell efficiency records versus the average cell efficiencies of commercial modules, Chap. 2)

> From this race, we learn that laboratory records for the current workhorse of photovoltaics, the silicon solar cells, preceded the average efficiencies of the module production with a typical delay of about ten years.

This time distance shows the still important potential providing a relative increase of commercial efficiencies of around 20 % to be expected within the next 10 years for the crystalline silicon technologies.

If we compare this with the newcomers, the thin-film solar cell technologies, the promising developments of recent years provide enough potential for efficiencies beyond 10 % and even up to 20 % to justify industrial efforts. It also shows that for all competitors, module efficiency remains a moving target.

Another important industrial goal is the speed of production or the specific yield for a given manufacturing capacity. It means, in general, how the ingredient materials are treated and processed in the module as the active layers of suitable performance.

For the cell technologies based on highly purified crystalline silicon, one important limitation is the time and energy required for the material to follow a certain physical path of heating and "freezing" in an appropriate manner. Today most techniques of silicon processing are volume-bound high-temperature heat transfer processes with inherent physical limitations in speed (see Chap. 1 for details). This holds for the step where the "raw material" is grown. The subsequent steps of the wafer processing technology are typically a throughput problem of yield and quality (e.g. sawing, diffusion/doping, cell connecting, modules, etc.).

In contrast to this actual crystalline silicon technology, the thin-film technologies in principle combine the module-related large-area manufacturing with the cell production from the beginning as an integrated process. So, for the thin-film manufacturing, there are the typical questions like:

- What is the optimum possible speed of deposition of material of sufficiently quality (efficiency) and for a sufficient large module area? Can the process be realized as a continuous in-line manufacturing?
- Which step becomes a bottleneck within a sequence of multiple-layer deposition steps necessarily interrupted by the multiple structuring to build connected cells in series within the integrated module?
- What are the cost structures for the complete manufacturing process concerning labor, materials and capital cost?

In this context it is the hope of the thin-film developers that their cost relations are more favorable to mass production than for wafer technology. In the end, highly automated and fully integrated production lines with low materials consumption together with low depreciation of production equipment may be achieved.

From these considerations the following boundary conditions and goals for the PV industry development may be derived:

Steady increase in technological quality with higher efficiencies (modules and system)

Reduction of specific manufacturing cost (per m^2 of module area as well as per kWh of system power output)
Improvements in materials and energy balance
Extension of existing market segments and development of new ones
Further evolution of export markets and global strategies

Next, the special industrial structures inherent to photovoltaics will be analyzed.

3.1.3 Industrial Structures

For the evaluation of the industries learning process, the structure of the industry itself also has to be considered. The industrial structures today reflect the learning story of the last decade, the dominant technology being based on (single- or poly-) crystalline silicon and wafers of typical size and thickness manufactured from electronic grade silicon sources.

Crystalline silicon is normally processed in separate steps, leading to subsequent different industries, for the production of

- Metallurgical grade silicon
- Highly purified electronic grade silicon
- Silicon crystallization and sawing into wafers
- Solar cell processing
- Module integration

followed by the system component manufacturers (inverters, batteries, other BOS-parts and services).

These manufacturing steps represent separate branches of the so-called solar photovoltaic industry, with quite different enterprises in size and competence (see Table 3.2).

Worldwide PV production adds up to about 30 enterprises for pure PV cell production (Fig. 3.1) and to about 50 companies including module production [IEA 2000]. In Germany only a few cell manufacturers exist (normally also manufacturing modules), together with a larger number of smaller companies aiming at specific markets for PV modules.

On the level of silicon raw material and wafer production, the number of enterprises is even smaller, and worldwide only about 10 companies cover 85 % of the capacity for wafers (Fig. 3.2).

The world PV industry meanwhile has passed the one billion dollar business threshold just at the turn of the century with encouraging growth rates. It has established considerable manufacturing capacities in a number of countries and is accepted as a prospering future global enterprise.

However, there is no guaranteed easy access to this future and the industries are still facing high risks. As was described in the first chapters, there is no general consensus about how to fill the gap between the present production cost level of PV and the cost level required for competitiveness in

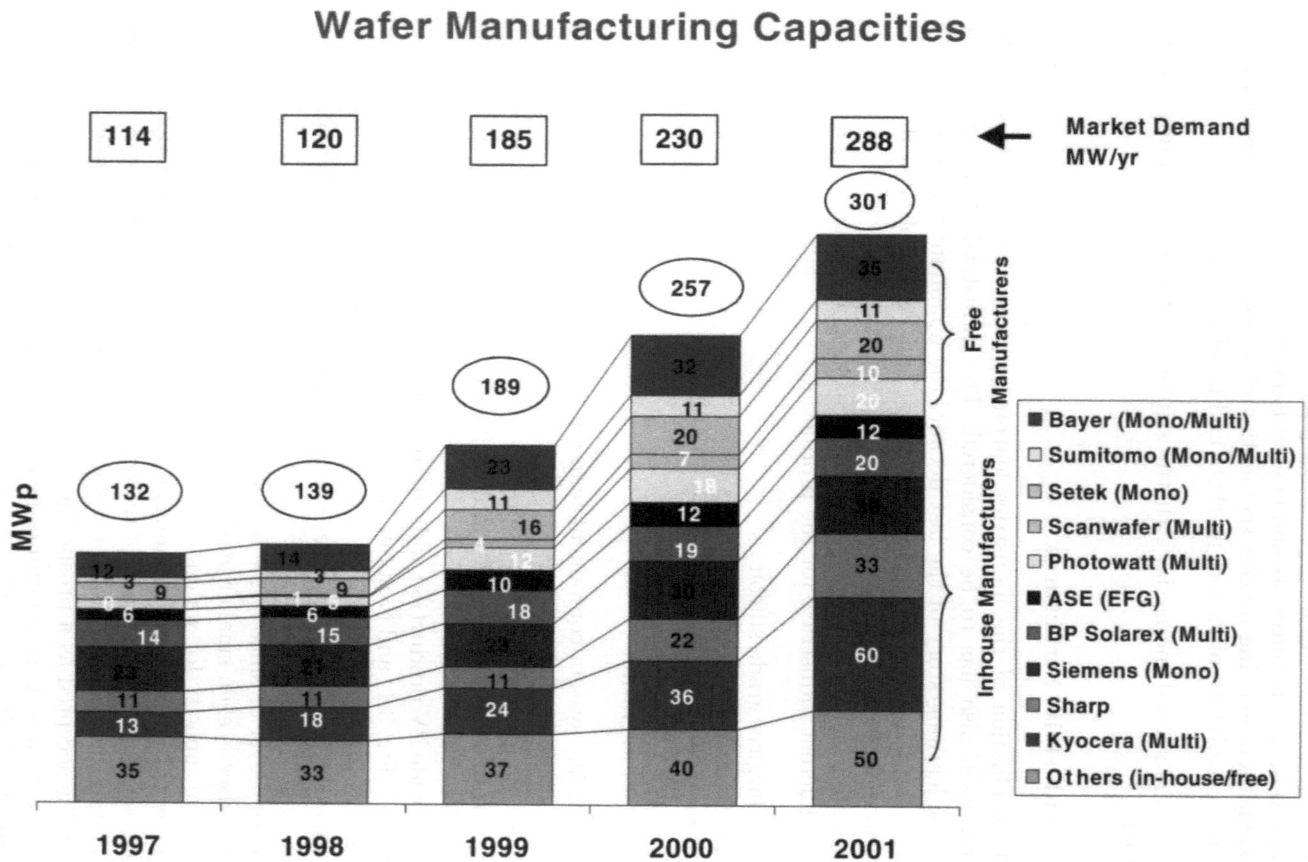

Fig. 3.1. History of industrial capacities of solar silicon wafer production (from [Woditsch 2000]) (graphic courtesy of Bayer)

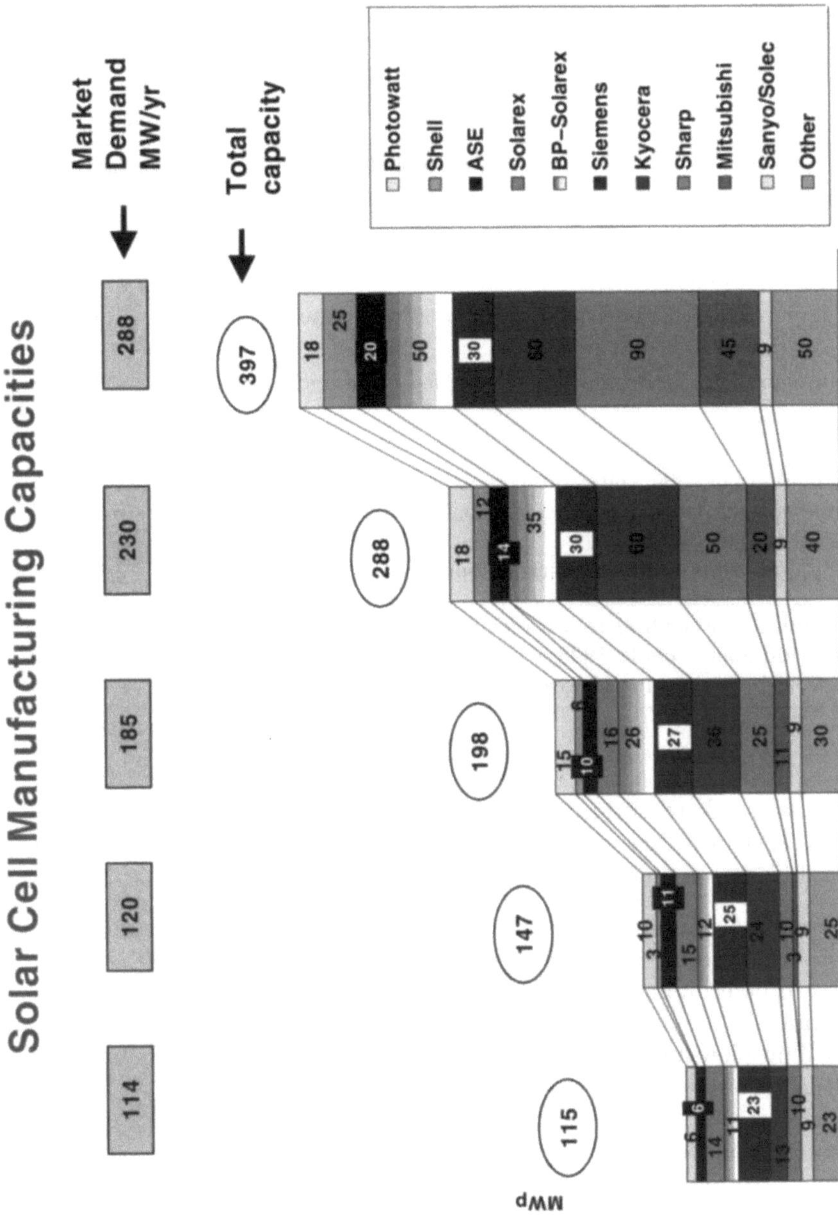

Fig. 3.2. Structure of today's photovoltaic Industry (graphic courtesy of Bayer [Woditsch 2000])

Table 3.2. Typical structure of PV industries

	Enterprises	
	World	(Germany)
Silicon feedstock	few global players	(≈ 500 t/yr)
Si wafer production	≈ 10	(1–2)
Solar cell fabrication	≈ 50	(≈ 5)
Module manufacturers	≈ 200	(≈ 20)
System component manufacturing	≈ 500 estimated	(≈ 50) (inverters: ≈ 20)
System suppliers and services	≈ few thousands	(≈ 800)

Dynamic:

- more mergers of solar cell producers expected
- enlargement of silicon feedstock capacities (only few fabricators, preferred chemical industry)
- rapidly increasing number of small suppliers and services

Figures estimated according to various publications and reports (e.g. [IEA 2000, PV-Produktion 2000, Schott 1998]). In general, it can be concluded, that Germany enjoys a 10 % share of industrial capacities.

everyday power supply, which is a scale of one order of magnitude whatever point of view is taken (corresponding to the a.m. goal of a reduction of the specific PV cost by a factor of ten within the next two decades).

The political assessments cover a wide range of statements:

- PV already has indigenous markets with proper growth in off-grid applications and doesn't need subsidies. Let the industry do the job until PV proves to be economic for further implementation in the general electricity supply.
- PV will be a key industry of the 21st century and deserves to be supported as such by, of course, necessarily high public investments followed up by commercial engagement as has been done for other key technologies of strategic importance.
- PV is a solution for decentralized electrification in developing countries and for some residual off-grid applications in industrial countries, and must first of all be started there.
- PV generators provide double output in the world's solar belt countries. It's there that business should be started.

and further statements of this kind. Rather visionary and optimistic positions are found for example, in [Scheer 1993], and rather pessimistic criticism, for example, in [Bradley 1997].

It is to be noted that today all global industrial competitors in USA, Europe, Japan etc. are undertaking substantial subsidies and promotion programs for PV. This is the more remarkable as PV today still ranges at the highest cost level of any form of new renewable energy technologies. Nevertheless, it seems that the arguments analyzed so far are promising, and that the long-term potential is convincing enough to justify the political efforts.

It is further remarkable that the subsidies given are strongly favoring grid-connected PV (see Chap. 1) where the gap between cost of PV generated electricity and the market price level of bulk electricity again is the highest possible.

The reason is that for reaching mass production, grid-connected systems greatly facilitate

- relatively standardized systems (few-kW systems) and attractive investment volume for suppliers
- obvious public assets
- preference of domestic investment and employment
- support of the domestic industries as a platform for later exports
- keeping global players of technology in the country by offering a guaranteed market

thus providing also benefit to political issues promoting domestic industries.

The global industrial player, however, only can be retained in a domestic market if there are advantages offered by national policies. On the other hand, typically the belief of investors in political decisions is not very strong. So investors prefer direct investment subsidies, providing reduced risk from the start. Instead, economists say that subsidies via electricity prices (as for instance the German feed-in law) provide a strong market pull with benefits concerning competition and system performance (as has been quite successfully demonstrated for wind power in recent years).

Therefore the decisions of industrial managers depend on their belief in medium-term effects of political programs and the stability of the political and economic context in a given marketplace for a scale of at least five to ten years. Furthermore, the discussion is strongly formed by the public debate on energy issues in general, where renewable energies are discussed in a ranking defined by other criteria (environment, energy dependence, nuclear debate, etc.).

In the following, before analyzing future prospects of PV industries, the way industries of this kind have been learning so far is discussed (compare also the excursion of the company ASE related to the history of the a-Si technology).

3.1.4 The Silicon Photovoltaic Industry

Behind the rather steady learning curve there is a process hidden including important structural developments sometimes taking place quite unsteadily concerning individual industrial enterprises, which only in total reflects a quasi-continuous development over longer periods.

For the later discussion of future strategies, it is important to analyze the way industry has been "learning" in the case of PV (for "learning" in research see Chap. 2).

Typically, learning during the first "terrestrial" phase of PV development (until around 1980) meant technological approaches of companies that mainly gained their experience from satellite space technology or as well from electronic grade silicon production. The standard of production literally was "manufacturing" of crystalline silicon solar cells.

The 1980s were dominated by efforts of big companies (even big oil companies) to industrialize those manufacturing processes to the level of mass production, with more or less success. Looking backward from now, the expectations during this first "boom" of terrestrial PV seem to have been rather optimistic if not unrealistic. These expectations, however, were largely triggered by the increase in US federal programs for the development of PV technology, which in turn were a response to a change in the long-term energy scenarios.

Even conservative authors and institutions promised PV a bright future (from [DeMeo 1984] it is quoted: "Good prospects for technical and for orderly business development point toward a bright future for PV as a source of bulk electric power" expected toward the turn of the century).

It can also be regarded as characteristic during that era that the market growth expected for the 1980s has been largely overestimated from a present point of view. Consequently, a number of early engaged big companies soon went under after a few years in the business as it became clear that there was no fast breakthrough to really cheap PV technologies just coming from the lab.

It is, however, interesting to compare the earlier market expectations and cost scenarios with the real historical learning curve the technology followed. Surprisingly enough (not surprising anyone who has a strong belief in learning curve theory) the goals fixed in early enthusiastic claims of a dawning PV century did not deviate much from the real learning curve. Instead they overestimated the speed of progress along this curve.

It remains a question whether such fast progress could have been managed at all. That such progress is not totally unrealistic can be seen from the market development of wind power, which in 2000 passed a 2 % share within the German public power supply balance, a result nobody anticipated a decade ago.

This first big industrial engagement during the 1980s and 1990s, however, also brought some clarification of how the process might go on, and that a steady learning effort over decades had to be faced to make photovoltaics a true factor in the global energy supply balance.

It is important to note that during these years progress was quite different in the individual branches of manufacturing given above. While processing of silicon wafers to solar cells was developed rapidly to lower cost and higher efficiencies, the primary production of solar silicon still depends on the availability of pure silicon as a byproduct of worldwide electronic grade silicon production. No substantial independent solar silicon production has been established yet, a fact that now turns out to become a major obstacle to the silicon cell strategy. This fact, however, is now realized to be a necessary next step in the development of an independent solar grade silicon industry, and a number of industrial projects have been announced recently.

The next step, the crystallization techniques, is covered by different industrial approaches of more or less maturity. A number of industrial processes derived from that first learning period are now commercially available for the production of solar grade silicon and also in use together with new competing processing technologies (e.g. the etch foil growth process of ASE Americas or the ribbon grown silicon resurrected in recent years).

The manufacturing of PV modules from silicon solar cells has developed as quite another level of industries, typically with much smaller companies and more oriented towards the diversities of the end user. While at the beginning of the 1980s practically all solar cell manufacturers were also producing modules, today the number of module manufacturers is distributed over a wide range of capacities, including a larger number of small companies exclusively producing modules (some only for specialized market niches). The latter companies are often also engaged as the last distributor of systems for the final customer. They mostly also offer further system components, either from own production or from supply industries.

For the workhorse of photovoltaic, the silicon technology, in Fig. 3.1 there are given the more important worldwide manufacturers, a dozen companies forming a capacity of production of silicon cells equivalent to about 384 MW_p. Compared with a worldwide shipment of 200 MW_p in 1999, this capacity has only been used with approximately 50 % of its nominal output. This situation has not been usual in photovoltaics. It resembles the now accelerating market, which forces investors to build capacities in due time together with the fact that the production facilities are often prototypes with longer startup times. It was also a fact, that during longer periods in the past, there were more module manufacturing capacities available than for cell production (modules standardized already in contrast to cell production).

All these capacities from an conservative industrial point of view still have to be considered more or less as pilot productions in the sense that their production equipment (except module lamination and framing where due to the larger number of small equipment also higher product numbers have been achieved) is individually designed and more or less not mass-produced equipment designed to standard specifications.

This is not unexpected, as the learning process so far fortunately seems to proceed faster than production. Practically none of these solar factories would be built a second time as a pure "blueprint copy," and specifications are continually modified (this experience is part of the valuable knowhow of any company and is kept rather secret!).

As for any fast-growing and knowhow intensive new industry, it must be kept in mind that the expected growth for the photovoltaic market (say around 20 % p.a., depending on nominal module power) produces a critical balance between up-scaling of manufacturing plant size and technology progress.

The balance is now well known for classical crystalline silicon technologies, being separated in quite different branches with inherent knowledge (like chemical industries for primary silicon or electronic industries for the cell device processes) and each with quite different optimal economic scale of production. This balance is not clear for the thin-film technologies yet.

3.1.5 Thin-Film Photovoltaic Technologies

For the new thin-film technologies, knowhow during the present pilot phase is still "borrowed" from other industrial branches, like the glass industry, the vacuum deposition facilities, etc.

For thin-film production, this knowhow then has to be integrated from the beginning into one optimal production unit. This means that such a company, to achieve the technical integration, at the same time also has to establish equipment and a staff of manpower, including a corresponding requested variety of engineering capabilities.

As the specific performance of a thin-film technology relies on a complex balance of material specifications together with processing "recipes," the whole fabrication could no longer be separated into different enterprises as in the case of silicon wafer-based module technologies.

A further argument regards the situation of PV companies concerning R&D goals. All solar photovoltaic companies today are subject to enormous pressure to qualify advanced technologies. Therefore, these "high tech" firms better could be characterized as R&D engineering companies with subsequent marketing rather than as production facilities with a supplementary R&D section (requesting only a few percent of the company's budget, which would be expected in the case of classical and mature technologies).

As a whole, the PV industry is a new and complex approach to a completely new product and market. This requires a well balanced development process where all steps of processing must be learned to avoid bottlenecks. Such "learning" induces greater investment risks, the larger the jump in each up-scaling. In classical chemical industries, for instance, up-scaling typically ranges in the maximum order of a factor of five and any higher up-scaling would be considered not to be compatible with safety and quality margins.

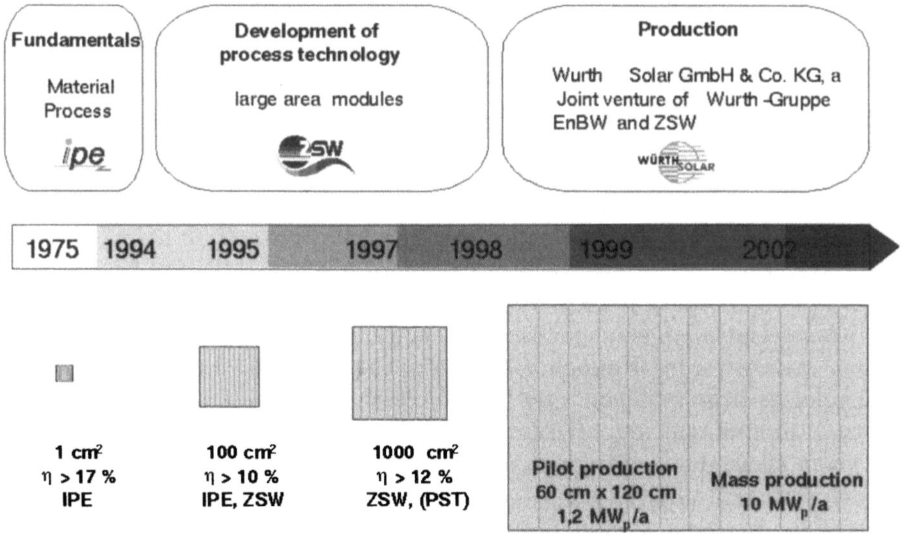

Fig. 3.3. One history of CIS-development

However, for thin-film photovoltaics, the up-scaling from laboratory to technical approvals and finally to pilot production has been and is still faster than for silicon technologies. A typical scenario, the historical route of the CIS technology of the Stuttgart group, may be cited (Fig. 3.3), covering a period of about 25 years. During the 1970s and 1980s it was governed by academic research and during the 1990s by technology-oriented qualification of production of complete large modules. The successful qualification then led to the joint venture Würth Solar with a 1.2 MW_p/yr pilot production facility now starting operation [Dimmler 2000].

This history fully reflects the conflict between the economy of scale required from the economic point of view and the technological risk of up-scaling in efficiency and module size within each step of learning.

From this, a fundamental difference between PV thin-film technologies and the electronic industries (a similarity valid for the classical silicon wafer technologies) has to be stressed.

Up-scaling in electronics took place with simultaneous (and even faster) miniaturization, such that the added value per unit physical hardware (a single crystalline silicon wafer) could be dramatically increased. In contrast to this, production in PV up-scaling is related to a physical dimension (the aperture area) which is not a pure production parameter but fixed by the application.

It has been analyzed ([Bossert 2000, Woodcook 1997]) that a reasonable break even for the thin-film technologies requires manufacturing capacities beyond 50 MW_p. On the other hand, real production experience was only existing for a-Si-technology (at a level of a few MW at most per manufacturing unit and scheduled capacities around 10 MW) while for the other technologies (CIS, CdTe) the 1 MW-level is now passed. All engaged companies at present are developing their strategy to deal with the gap between the actual knowhow and the envisaged multi-MW production at minimal investment risk.

The consequences and measures taken turn out to be different. For instance, in the case of CIS, the companies Siemens Solar and Würth Solar [Powalla 2000] have decided as a next step to confirm their technologies at the 1 MW production level before scaling up to the 10 MW level, while ANTEC went directly to a 10 MW fabrication [Bonnet 1998].

Actual and planned capacities for the near future for solar cell manufacturing were given by [Rannels 2000] up to the year 2000 separating thin-film technologies from c-Silicon, and by [Woditsch 2000] up to the year 2001 separating the different manufacturers.

From these data and from earlier sources [Johnson 1995] it can be concluded that typically the capacity factor of solar cell manufacturing plants is fluctuating between around 70 and 90 %, reflecting the difficulty of investors to decide upon a new factory according to market prospects. At present there exists a tendency to optimism, which in turn leads to faster growth of planned capacities than actual market growth.

To be fair, it must be stated that in the last two decades the growth of manufacturing capacities sometimes has been dramatically overestimated concerning both the whole market and the development of typical manufacturing unit sizes.

From [Rannels 2000] it can also be derived that in 1998 there was high expectation concerning new thin-film plant capacities adding up to about 155 MW_p production capacities for the three thin-film technologies considered mature aSi, CIS (incl. CIGS), and CdTe. In total, these technologies are covering 32.5 % of a total capacity of 473 MW_p expected for the year 2000. This, obviously, has been somewhat overestimated according to actual figures [Woditsch 2000] quoting for the year 2001 a total capacity of 384 MW_p of crystalline and polycrystalline silicon solar cells. It seems that in 1998 the expectations of immediate breakthrough of thin-film technologies at capacities of industrial scale were too optimistic.

On the other hand, the fabric for c-Si-solar cell manufacturing seems to have been realized in due time, in good agreement with the capacities planned in 1998. This is not surprising as the c-Si solar PV industries are now well established and already in a steady state of development compared to the yet unsteady thin-film development.

The only thin-film technology which has played a continuous and substantial role so far in the PV market is amorphous silicon technology (compare also the excursion chapter of the story of ASE/Phototronics "Experiences and Lessons from a Pilot Line" on page 170). Historically having already been several times considered a major alternative to wafer-based silicon technologies, amorphous silicon obviously still faces a great barrier towards higher market shares with relatively low average efficiencies around 6 to 8 %. As there is hope that higher efficiencies around 10 % can be achieved in mass production with more sophisticated multiple solar cell structures (in the laboratory combinations of amorphous and micro-crystalline structures also form new thin-film options for the future with higher efficiencies), the future of a-Si-cells is still open. At present, the steady transformation of pure a-Si concepts towards hybrid concepts combining a-Si with micro-crystalline multiple cell concepts is a reason to keep with amorphous silicon cell concepts.

Recently published evaluations [Bossert 2000] of a-Si technologies (including the potential of variations as combinations of a-Si with thin-film crystalline silicon cells) come to the conclusion that the future potential even could outstrip the alternative "classical" CIS, CdTe thin-film technologie now entering the stage.

The fundamental situation, however, seems to be different for the CIS and CdTe thin-film technologies. There, laboratory records of efficiencies (achieved with no higher or sophisticated cell concepts) now rapidly approaching the 20 %-threshold provide enough potential for future improvements. If a similar rule (a ten-year time delay between laboratory results and their implementation in production) holds for thin-film technology transfer, then efficiencies of around 15 % might be expected for 2010 in large scale production.

Today manufacturing at a pilot stage of around 1 MW series production (which means around 10 000 modules per year) with anticipated average efficiencies around 10 % will approve the manufacturing concepts within the next one to two years.

On the other hand, it must also be concluded that it still takes another couple of years of learning at this pilot stage, before the risk of larger investments could be afforded. This means, that the 10 MW class for the new thin-film technologies will be entered within the next five years.

To place a new technology in competition on the market, the size of production defines the economy of scale and therefore the competitiveness, not only with the already established technologies but also with any competitor in the specific new technology. So, again a compromise must be found between technological up-scaling risks and economic up-scaling advantages.

If we suppose an optimistic market expectation for the next years (growth rates 20 % and more), a size of a 10 MW factory seems a good compromise, provided that a sound technical basis is given for all production components. If we assume a 50 to 70 % capacity factor for the beginning, that means that even such a modest approach from the beginning must clear a 5 % bar of market share to place a new label within such a short time.

The above arguments define a certain speed of progress, harmonizing technology advancement with industrial investment strategies.

Finally, what would a realistic time scale for learning be? Even at the pilot production level (by definition intended as a "learning" investment and purpose), a couple of years of operation is necessary to gain reliable experience. If this experience must be available in time for the planning of a next generation production line, then at least three to four years between the subsequent pilot investment decisions is required (at least one year procurement and installation, one year for commissioning and start up, 1 year to establish routine production).

From these arguments, the following conclusions can be drawn:

- Innovations at industrial level follow certain stages of experience which must be arranged with investment cycles (typically 3–5 years). Such gain of production experience and quality cannot be accelerated without limit and no investor would take the risk of forgoing this gain. So the investor will decide upon the next up-scaling only after a certain time.
- Learning by production is a difficult compromise between technology risks and economic market "pull," limiting the speed of progress.
- During a technology-driven period of high risk (which PV went through so far) the capacities of fabrication could only be increased with a speed beyond the natural progress of technological learning by increasing pilot capacities. This can happen by increasing the number of independent players, or by public subsidies reducing risks for the individual enterprise.
- Technological learning of an individual company is realized in sequential stages and not by a continuous process. This makes the individual investment rather difficult expecially concerning the date of decision. The risk remains of cutting further operational experience with a running production to the benefit of a cost reduction by earlier up-scaling.

Up to now only the learning process for the "product" PV module has been taken as relevant for the progress. However, the real product of PV is a complete system for a specific application, including more than just modules. To estimate the cost reduction potential and the relevant learning strategies there must also be considered the complete economic balance sheet of investment and production. This will be done next.

3.2 Cost Structures and Reduction Potential

3.2.1 Crystalline Silicon Modules

Solar PV generator systems can be separated both in principle and practice into two main parts, the PV module and the "rest" of the components required to form a system fulfilling a certain purpose or power supply for a specific demand.

It is one of the inherent big advantages of photovoltaics, that the solar cell principle allows a maximum of flexibility of true modular design of PV modules over an extremely wide range of rated power not known for any other power technology. This leads to a strong standard of module design from the beginning.

Furthermore PV modules allow a rather high degree of standardization as a mass product (a setup of typical ratings of power/voltage/current), which is of high value for module fabrication. The same does not hold for the system components in general, but only for selected applications, e.g. for grid-connected generators.

We consider first the cost structure of PV modules with respect to their industrial manufacturing steps, according to the results of a number of recent comparative studies ([Beneking 1997, Bonnet 1998, Bossert 2000, Bruton 1996, Francis 2000, Fuhs 1998, KPMG 1999, Kraus 1999, ASE 2000, Maycock 1997, Powalla 2000, Woditsch 2000, Woodcook 1997, Zweibel 1999]).

3.2.2 The Crystalline Silicon Case

The cost structure of silicon solar cell-module production has been followed up for many years. During these years the typical design of a PV module has not been showing substantial variation, which is quite surprising in view of the progress in all those years. PV modules are built after all from Si-wafers of almost uniform size (around 10–15 cm diameter) and only little variation in shape. They are still connected in series to achieve a module operating voltage level for further string connections, building larger solar generators. The solar cell arrangement is then laminated between two glass plates (framed or frameless).

Furthermore, the separation of industrial manufacturing steps in different companies also brought economic clearness, cost data have been available explicitly at every step.

It is therefore possible and a reasonable approach to pursue the development of the contributions of the building components of such classical PV modules concerning materials, labor, and capital cost.

In this context it becomes important to follow the cost structures at the consecutive levels of production, where the output at one step of production will be the material input and cost factor input at the next stage. For instance, the Si wafer is a material cost factor for the solar cell factory while the production of the wafer again includes material input (electronic grade silicon raw material) together with additional labor and capital cost.

Summarizing this chain of production steps, one ends up with a matrix representing the steps of production and allowing to add up the total cost apportioned to real primary material, total labor and total capital cost for manufacturing equipment. In Table 3.3 the development scenario for a best guess of today's technology progress in the near future is derived. It is interesting that the typical size of manufacturing capacity is assumed to increase to volumes five times the largest capacity of today (from 20 to 100 MW_p)

Table 3.3. Scenario for polycrystalline silicon PV modules (cost ex factory in DM/W_p)

Cost Share	Status 2000	Plan 2005	Expected 2010
Si wafer	2.75	2.00	1.55
cell	1.90	1.20	0.90
module	1.90	1.30	1.00
total	6.55	4.50	3.45
Assumptions			
efficiency (%)	13.5	16	17
wafer thickness (μm)	330	300	270
raw silicon (DM/kg)	40	30	25
yield (%)	85	90	95
production capacity (MW)	20	60	100

within the next ten years. All other parameters are only variable within small margins, e.g. average efficiency from 13.5 to 17 %. It is anticipated that this increase in production and specific technology improvements will lead to a reduction of production cost by a factor of two.

For silicon technology it remains clear that progress in economy is largely dominated by scaling effects in production volume. This is due to the fact that module construction is principally related to the wafer as an elementary building block of photovoltaics, which has also been optimized over a long period of several decades. Wafer technology also took important benefits from the general progress in crystalline silicon wafer-based electronic technology.

There are ways beyond the single wafer, as for instance ribbon-grown silicon (RGS) and etch foil-grown sheet (EFG), and others which have been developed beyond electronic as independent and pure solar technologies. These technologies in principle start with production of large-area polycrystalline silicon sheets. The sheets, however, are then cut to wafers again to supply the usual cell manufacturers. It is not clear at present how these technologies can be successfully used for direct large- area modules (apart from problems of homogeneous surface quality or structuring into cell assembles). So these are alternatives to the classical wafer cutting process with improved process parameters (like fewer cutting losses and fewer steps) and economic advantages. In [ASE 2000] the main advantage of EFG-material over classical wafer technology is derived from significantly reduced demand for material which is assumed also to hold in the long run.

The total chain of all manufacturing steps from the silicon to wafer, to cell and module integration is taken from [Francis 2000] in an overview (Table 3.4)

Table 3.4. Total cost structure matrix of cast polycrystalline silicon wafer-based PV module manufacturing (from [Francis 2000])

	Specific cost share (US$/Wp)[a]				
	equipment depreciation	material	labor	yield losses	total
Si wafer	0.10	0.57	0.11	0.04	0.82
Cell	0.14	0.08	0.16	0.06	0.44
Module	0.04	0.52	0.21	0.09	0.86
Total	0.28	1.17	0.48	0.19	2.12

[a] 1 US$ = 1.09 Euro (Feb. 2001)

adding up all material, labor and investment cost (depreciation) employed in all three steps of manufacturing.

In [Francis 2000] the comparative analysis has been completed for a 10 MW plant capacity including labor, materials and depreciation at all three levels of wafer technology. The resulting cost figures are somewhat different from the above-mentioned analysis in [Woditsch 2000], which were related to a 20 MW production unit.

However, the picture is coherent in the principal aspects if one takes into account that the economic parameters of any investment calculation are also subject to a number of quite variable parameters (specific cost of labor, depreciation, interest rates currency exchange rates etc.) which may explain such deviations.

The existing and planned manufacturing capacities based on the three workhorse technologies of crystalline wafer-based photovoltaic are given in the Table 3.5 [Woditsch 2000].

Si wafer production represents a main cost factor for c-Si modules (around 40 %), being of strategic character (at present the material input is extremely dependent on the electronic grade silicon market with rising cost share, the pure silicon material input being around 25 % of the total module cost).

It is therefore of interest to look for the cost scenario of the wafer production. From [Woditsch 2000] a cost breaking can be quoted, indicating the measures to be taken for further cost reduction (Fig. 3.4).

In this context, the alternative wafer production techniques show remarkable differences. In [ASE 2000] a cost comparison between wafer production from cast polycristalline silicon with an edge foil grown (EFG) wafer production is given, promising a larger cost reduction potential and a higher efficiency of material utilization for the EFG technology (see Table 3.6).

For the recently revived ribbon-grown technologies (RGS), no proven figures at a larger production scale are available.

Table 3.5. Crystalline silicon solar cell manufacturers [Woditsch 2000]

Crystalline Technology	Capacity (2000) (MW)	Planned capacity (2001/02) (MW)	Country
Monocrystalline Si			
BP Solarex	20	10	SP, IND, AUS
Isofoton	8	8	ESP
Sharp	10		JP
Siemens Solar	25	5	US
Solec International	10	10	US
Total	73	33	
Polycrystalline Si			
ASE	6	6	GE
BP Solarex	20	10	FR
Ersol	2	8	GE
Kyocera	36	20–30	JP
Mitsubishi Electric	10	12	JP
Photowatt	20		FR
Sharp	44	22	JP
Shell	12	25	GE
Total	150	103–113	
Sheet/edge grown Si			
ASE Americas	4		US
ASE GmbH	6		GE
Astropower	10	10	US
Ebara Solar	pilot production	pilot production	US
Evergreen Solar	pilot production	pilot production	US
Total	20	20	
Total crystalline technologies	243	156–166	Total capacity ≈ 400

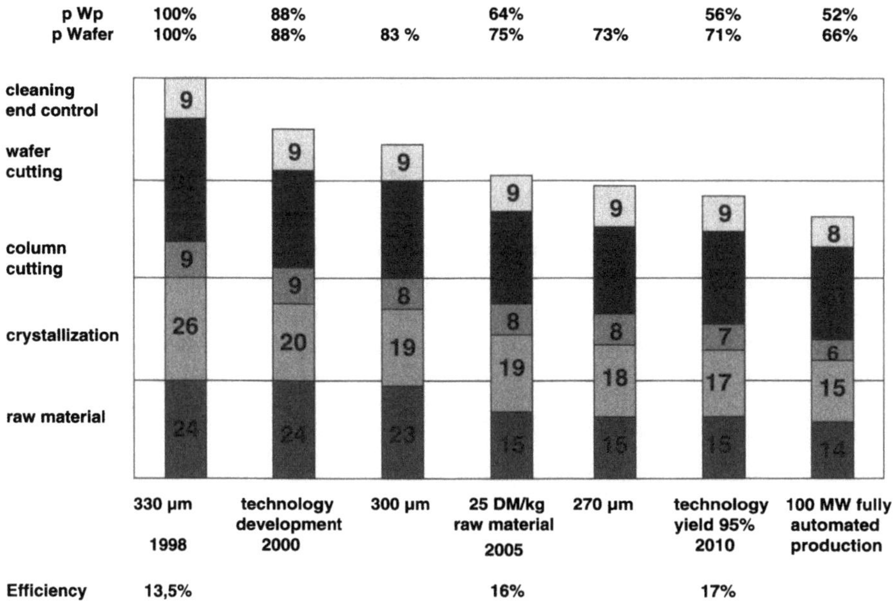

Fig. 3.4. Cost reduction potential of silicon wafer production for solar cells.

Table 3.6. Long-term utilization of polycrystalline silicon material in various wafer production technologies (after [ASE 2000])

	mc-Si		EFG	
	status	long-term	status	long-term
Input[a] (g/W_p)	11.4	4.0	5.9	1.6
Fraction used	46 %	37 %	85 %	95 %
Wafer (g/Wp)	5.2	1.5	5.0	1.5
Wafer thickness (µm)	330	100	300	100
Cell efficiency (%)	14.7	16.5	14.0	16.0
Specific[a] cost ($/$W_p$)	0.23	0.08	0.12	0.03

[a] assumed 20 US$/kg Si

Considering the total balances of materials, labor, and capital cost, the latter also in principle opens another level of the same implemented qualities usually hidden behind the calculation of investment cost respectively the depreciation (of production equipment). Here the value depends on economic parameters as depreciation periods and added value of complex machinery.

Finally, all these factors can be summarized by a so-called total life cycle analysis of energy and material required for a final purpose like electricity supply for some consumer or for feeding into the grid, in relation to the total input of raw materials, labor, and capital.

While silicon solar cells represent a typical classical structure in this context, the situation is different and new for thin-film technologies in general.

3.2.3 Thin-Film Technologies

For thin-film technologies, attempts at industrial qualification go back to the early 1980s and have been in the focus of interest since then (for instance [DeMeo 1984]). Since then a large number of candidates of principally photovoltaic active semiconductor materials have been investigated. Up to now, three of them

Amorphous silicon (a-Si)
Cadmium Telluride (CdT)
Copper Indium Diselenide (CIS)

have shown breakthroughs from the laboratory scale to a serious pilot production level, providing sufficient experience and technical information to evaluate their future economic potential.

Recognizing them as real candidates now, an increasing number of comparative studies for the evaluation of thin-film technologies have been carried out recently ([Bruton 1996, Woodcook 1997, NCPV 1999, Kraus 1999, Francis 2000, Bossert 2000, Woditsch 2000]).

The main physical difference between Si wafer-based technologies and the thin-film technologies is the ratio of the true light-absorbing semiconductor material, which is deposited as a thin layer a few microns thick, to the rest of material, especially the bulk material of the module, typically glass. While for silicon wafers this thickness ration is of the order 1:10, for the thin-film modules it is of the order of 1:1 000.

This means that the material balance in the primary production of a thin-film PV module is dominated by the substrate, which defines another limiting cost situation for future PV mass production. It also leads the way to highly integrated PV in a general infrastructure, where the remaining (thick) substrate is now ruled by other tasks beyond energy (architectural, protection, esthetic, etc).

As a second fundamental difference between crystalline silicon and thin-film solar modules, the chemical compounds and the processes of deposition or of secondary chemical reactions to form the active semiconductor sandwich structures on the surface of a substrate are interconnected and more complex. The structuring and semiconductor formation of the solar cell is no longer separated from the crystallization process. This means that the thin-film technologies form another class of fabrication complex.

According to the integrated processing, thin-film technologies have been developed from the beginning as an integrated manufacturing process in one fabrication step instead of five typical steps for c-Si modules (raw silicon, purified electronic grade silicon, crystallization and cutting (wafer), cell production, module integration), historically forming quite different manufacturing branches.

This also means that such an integrated thin-film module production scheme enforces for a given size of production a simultaneous optimization of all intermediate steps for the same size. In contrast to this, the technical and economical optimization of the separated subsequent steps of c-Si module production allows an independent economic production with respect to investment conditions and market prospects (a qualified personal view may be taken from [Martin 2001], where all the uncertainties of an investor with relation to the investment decision are clearly stated).

All this leads to the fact that the cost projections of thin-film module production follow another route. Together with the fact that the processing of large-area substrates (glass being the material dominance of the substrate) is a strong economic boundary condition (around 10 000 cm^2 of glass plates compared to silicon wafers of only 100 cm^2), thin-films also require largely integrated and automatized production. It is the hope that photovoltaics in the far future might be managed in a way similar to the glass processing industry, which annually produces millions of sqare meters of flat glass of high functional qualities by complex surface treatment, with specific cost as low as 10 Euro/m^2.

Finally, the most important difference, of course, is the complete processing of the photovoltaic module in one fabrication starting only with primary (metallurgical grade) material ingredients as provided by the chemical industry, together with substrate materials of standardized quality (glass, steel, etc.). It is this vision of completely merged manufacturing that made thin-film solar cells attractive candidates since the early times of PV research.

Economically, starting from the beginning with larger substrates means higher risks at later steps of the fabrication with increasing added value. Therefore, the quality control and management is an essential requirement for this integrated fabrication.

The field of optimization and improvement of thin-film technologies therefore follows a route different from silicon. There are two main parameters as measures of progress, the area of the integrated module (larger modules promise better economy of equipment) and the quality of the thin-film cell, with a main parameter, the overall solar cell efficiency.

During the historical progress in c-Si cell design the wafer size has not been varied very much, and the main focus of improvement has always been the cell efficiency.

In contrast to this, thin-film cell improvement always had to match as well enlargement of the cell (or module) area simultaneously with increase of efficiency (optimization in a two-parameter space).

This fact also dominates the physical and economic race of thin-film technology in industrial scales. It became very important to define reasonable steps of how to maximize the speed of progress by optimizing

- Efficiency
- Module size
- Production scale

Module size is now becoming one of the most important parameters, in contrast to the classical silicon wafer technology, where the module could be designed in a free manner at the end of cell processing.

For thin-film technologies, economic improvement is now dominated by the increase in module size. Here, the parameter of scale is actually in rapid progress with even three orders of magnitude, if we consider the laboratory scale (cm^2), the pilot production (typically 100 to 1 000 cm^2) up to the typical module size of about 10 000 cm^2.

None of the candidates mentioned above, however, has cleared this bar yet to full economical large scale production.

Also, the so-called integrated manufacturing of the structuring into cells and their electrical serial connection is totally different form silicon cell technology. There the challenge is defined by the repeated structuring of large modules with high precision in the micrometer scale at high speed (mechanical, laser structuring, etc.). While silicon wafer modules may be integrated by direct manual labor with solar cells supplied from elsewhere, thin-film modules must be processed in one line, which has to be optimized as a holistic manufacturing process.

It must not surprise under these circumstances that the route of development for thin-film modules and its up-scaling went different ways and the progress is to be described in another context, namely the Combined evolution of module size, efficiency and process, speed together characterizing a complex technological race.

As an example of efficiency versus module-size development analysis, again the case of CIS technology may be quoted. From [Dimmler 2000] and [Powalla 2000] actual results can be found (see Table 3.7).

Table 3.7. Best efficiencies of CIS module versus module size

Aperture area (cm^2)	Number of cells	Efficiency (%)	Output power (W)	Year
91	15	13.9	1.3	1996
60	12	12.7	0.8	1997
0.5	1	16.1	0.008	1997
718	41	11.5	8.2	1998
717	41	9.7	7	1998
731	53	12.8	9.3	1999

The development of large thin-film solar modules must still be considered a risky process now involving large investments, because further "learning" can only be achieved at industrial production scale. A typical threshold of competitiveness for the next stage can be formulated as:

> Qualification of a thin-film technology with at least 10 % module efficiency at module sizes around 1 m^2 and at production cost around 1 Euro/W_p, which means a 10 MW-production capacity.

3.2.4 Cost Scenario for 2010 and Beyond

Reliable data of cost structures for thin-film solar modules can so far only be derived from pilot production experience (if data are available and certifiable at all). The results recently summarized in [Francis 2000] can be taken as a picture of the status of thin-film technologies and their potential in comparison to c-Si technologies (see Table 3.8).

From an overall point of view, the cost level seems to be quite comparable for the different technologies, with material cost somewhat higher for the c-Si technology (55–60 %) whereas the thin-film technologies (also not surprisingly) show typically higher fixed-cost shares. For thin-film technologies in general, low production yields have been supposed, which is justified in view of the still pilot production character. In [Francis 2000] it has also been stated that thin-film technologies have the potential to realize lower cost mainly by up-scaling of production volume and improvement of yield.

The conclusions of all cost studies differ in the details, but in general agree that the break-even of thin-film solar modules with silicon wafer-cell technology will be reached beyond the 10 MW-production capacity range (nearer to the 50 MW-level) at a cost level around 0.5 to 1 EURO per W_p. It is expected that this goal will be reached within this decade, by 2010.

Table 3.8. Direct manufacturing cost for PV modules in 2000 (10 MW-capacity) ([Francis 2000]) (W_p).

	CZ Si	Cast Si	a-Si	Cd Te	CIS
depreciation cost	0.25	0.28	0.44	0.28	0.26
yield loss	0.24	0.17	0.40	0.45	0.44
labor	0.45	0.48	0.70	0.52	0.52
materials	1.51	1.19	1.16	1.05	1.03
total	2.45	2.12	2.70	2.30	2.25
efficiency (stable/average)	15 %	14 %	6 %	8 %	9 %
yield	85 %	85 %	75 %	65 %	65 %

The higher dependence on up-scaling is not unexpected due to the larger potential in automatized production. In the end, it will be this automatic production together with a drastic lower demand of material that will define the advantage of thin-film technologies over wafer technology.

The comparison between the present PV workhorse, crystalline silicon technology, so far only includes the flat plate module technique, where glass is the elementary substrate (back and front), guaranteeing high uniformity of the final product. This is not surprising, as new technologies first must survive competition with existing ones.

According to [Francis 2000], all compared technologies show a similar potential of cost reduction by almost 50 % within the next ten years, mainly by up-scaling (from the 10 MW level up to the 100 MW level) and by improving yield, together with moderate increase in module efficiency. So for the next ten years until 2010, the proposed cost reduction factor may be well within range of the present industrial structures and abilities (see Table 3.9).

But what about the scenario beyond 2010 for the various technologies?

A true long-term approximation for thin-film technologies has been discussed in [Zweibel 1999], calculating a limiting cost level of around $50 US per m^2 of a large module (still based on a conventional concept of flat plate glass substrate, etc.), based on detailed analysis of materials demand and production techniques. The real long-term manufacturing capacity has not been fixed in this study, but should be assumed in the range of 100 MW at least.

The figures of this study can be compared with results from recent studies based on specific production schemes ([Powalla 2000, Bruton 1996]) (see Table 3.10).

From the long-term aspects, it is interesting to note that the material cost becomes dominant in the cost balance, whereas (somewhat unexpectedly) the equipment vs. investment costs are not dominant. This gives hope that large-scale production will be realizable from an investor's point of view. On

Table 3.9. Cost reduction scenario for specific module production by 2010 (W_p)

	Direct manufacturing cost		Main measures taken		
Technology	2000	2010	Av. efficiency (%)	Yield (%)	Capacity (MW)
CZ-Si	2.45	1.45	15 ⟶ 18	85 ⟶ 95	10 ⟶ 100
Cast-Si	2.12	1.15	14 ⟶ 17	85 ⟶ 95	10 ⟶ 100
a-Si	2.70	1.40	6 ⟶ 8	75 ⟶ 90	10 ⟶ 100
Cd Te	2.30	0.95	8 ⟶ 12	65 ⟶ 90	10 ⟶ 100
CIS	2.25	1.00	9 ⟶ 13	65 ⟶ 90	10 ⟶ 100

Table 3.10. Specific manufacturing cost for large-area CIS modules

Reference	ZSW[a]		APAS[b]	Long-Term[c]
Capacity (MW_p/a)	1.2 (pilot)	10.1	60	Large-scale
Cost (DM/W_p)	6.85	2.30	1.00	0.3 ($/$W_p$)
Cost share				
Capital cost	32 %	30 %	28 %	9 %
Material	18 %	20 %	42 %	58 %
Labor	36 %	30 %	16 %	12 %
Other	14 %	20 %	14 %	21 %

[a] ZSW internal study 1996
[b] [Bruton 1996]
[c] [Zweibel 1999]

the other hand, this picture makes clear that for large production of thin-film modules, the cost of the material used will be of high relevance. The technical evolution will then enforce further research for thinner cell structures and higher efficiencies.

On the far horizon, thin-film technologies in principle allow far more revolutionary applications of photovoltaics compared to the silicon-wafer technologies. This is due to the fact that only thin-film solar cells allow the use of true surface-adapted or integrated technological approaches.

The next generation of thin-film solar cells, already in the laboratories, will bring solutions on flexible substrates (foils, metallic, organic etc.) as well as the deposition of PV active thin-films on arbitrary geometries and surfaces.

This development will induce then again a revolution in manufacturing, because the added value of a PV activated surface can only be optimized by fully integrating the manufacturing of PV with the conventional product of the appliance (architecture, mobile equipment, roof tails, etc.).

As soon as thin-film deposition and structuring of a flexible substrate or non-flat geometry becomes an accessible technology, a much-enlarged variety of new products and markets will be created.

Qualitatively, PV will conquer new areas of unique applications not imagined before and not realizable by any other power technology.

Consequently the complete system technology also has to be redefined to match such new applications, where PV and appliances are highly integrated, demanding suitable power inverters, storage devices, and control intelligence.

For these new applications, reliable economic data cannot be defined yet and the technolgies are subject to the fantasy of the engineer and the designer.

Apart from these long-term goals within a much wider scope of photovoltaic applications, the fast up-scaling of production capacities in total (following the anticipated market growth of about 20 % per year and more) as well as in individual fabrication capacity is a great challenge.

Whereas existing c-Si technologies start from a well experienced status, the new thin-film candidates have to join the process of up-scaling by factors of one order of magnitude almost from laboratory experience. An exception may be amorphous silicon, where the 10 MW-level is already realized.

Experience with the 10 MW level must be gained in time to match the next stage of competition in the range of a 100 MW-investment already visible on the horizon. Such fast up-scaling is *not* usual in industry and also did *not* take place in the history of photovoltaic.

A lesson of the risks implied can be taken from the long story of amorphous silicon, which repeatedly had been proclaimed the winner of the race towards large-scale photovoltaic thin-film production. So, probably a 100 MW-investment with classical c-Si technologies will be realized earlier than for the new thin-film technologies.

For the political evaluation of the technologies, one comes to different conclusions according to the degree of realization and reliability of technological experience which one assigns to the respective production technique. For instance, despite the deficits in module efficiency, some studies favor a-Si technology for large-scale production (see e.g. a recent analysis on behalf of the Netherlands government [Bossert 2000] recommending national engagement in a-Si).

The evaluation of thin-film technologies at different stages of their development often leads to a variable ranking. This must also be expected for the future, and there will probably no single winner in this race.

A simple argument supports the conclusion to keep the race open. There will always remain inherent advantages and disadvantages of any material choice for PV cells, leading to a different weight of properties according to the application chosen. This will also lead to a separation of technologies into different market segments in the future. It therefore makes real sense to keep all of them in a competition.

3.3 System Approach

The system components or so called balance of system (BOS) components of a "conventional" PV application may be classified into typical parts:

- Inverters
- DC wiring and connections
- Controller and operating units
- Mounting structures
- Safety components

To these hardware components further non-hardware cost shares have to be added, like planning cost, transport, installation and marketing. One should be careful not to underestimate the importance of the latter costs. If photovoltaics are to be sold at a much lower cost compared to today, it becomes an important question whether these "non-hardware" costs (one could also call them organizational and logistic costs) can be reduced with the same down-scaling rules as assumed for the basic hardware components. We will return to this point later.

For a typical actual grid-connected PV generator with two respectively 100 kW nominal rated power, the system shows the following cost breakdown and electricity generation cost, related to a typical location in Germany (with 1 000 kWh per m^2 and year of solar radiation; in southern sun-rich locations in Europe, generation costs would easily be reduced by a factor of almost two!) (see Table 3.11):

The cost levels are quite typical, but also show a larger statistical variation according to different market conditions. In general, one must state that PV-generated electricity is at least a factor of 5–10 more expensive than grid-supplied electricity (even neglecting the debate about capacity credit).

According to [IEA 2000], the bandwidth of investment cost ranges from \$6 to \$12 US per kW_p for 1–3 kW_p-systems, mainly due to various investment conditions like interest rates (the latter is very important, as the operational costs of PV are rather small compared to the investment cost). The costs also include a certain depreciation period, which depends on physical lifetime together with the capital payback time. None of the rather new existing PV

Table 3.11. Economy of typical grid-connected PV systems (status 2000) ([Staiß 2000])

		Roof system	Free site
Installed nominal power	kW_p	2	100
Global radiation	$kWh/(m^2yr)$	1 000	1 000
Total radiation on module aperture	$kWh/(m^2yr)$	1 120	1 120
Module efficiency	%	13	13
Yield (AC)	kWh/yr	1 600	85 000
Depreciation time	yr	25 (20)	20
Investment	DM[a]	28 000	1 000 000
Specific investment	DM/kW_p	14 000	10 000
Interest rate	%	6	6
Capital cost	$DM/(kW_p\ yr)$	1 095 (1 220)	436
Operational cost	$DM/(kW_p\ yr)$	75	50
Electricity generation cost	DM/kWh	1.46 (1.62)	1.08

[a] 1 DM ≈ 0,51 Euro, excluding VAT

Table 3.12. BOS-cost breakdown of grid-connected small PV systems (status 2000)

Installed nom. Power	kW_p	1	5	20
Specific investment	DM/W_p	16.7	12.7	10.75
Cost sharing	%			
Modules		51	58	65
Inverter		14	15	15
Mounting		7	6	8
Cabling		9	9	5
Planning and installation		19	12	7

generators has reached this age, and assumptions are speculative. Typically a depreciation of 20 years is common, leading to capital costs of around 1 000 DM per kW_p and year, which is more than ten times the operational cost.

The general trend of cost reduction with the size of a power-supply system also holds for PV technology, but with one important difference compared to conventional technologies (e.g., see cost breakdown quoted from [Meinhardt 2000] (see Table 3.12).

The part of the PV modules to be integrated into a generator represents in itself a highly modular design, whose specific cost is largely independent of the technical size of the generator (apart from a price reduction for larger orders).

The rest of the system, the BOS part, follows different rules.

For a small generator, the cost of service for planning and installation today cover almost 20 % of the total investment, and only half of it for larger generators. Furthermore, depreciation of equipment is to be calculated differently for the modules compared to more conventional components like power electronic. Despite the fact that a PV generator is a long-term investment with 20 to 25 years expected lifetime (a guarantee already given for the modules), the total life time cost still contains some risks.

3.3.1 Cost Reduction Potential of Balance of System Components

From this picture it becomes clear that for the long-term potential of cost reduction one has to consider the BOS parts of a PV system in detail. Since the beginning of the industrial period of photovoltaics, it has always been doubted whether photovoltaics could become economic due to the high specific BOS cost. The skepticism was sometimes formulated in the drastic way "Even if solar cells were free, the rest of the system would always remain too expensive".

The fact that a large cost share is really of "modular" character, on the one hand, represents a great advantage of PV technology for widespread applications. On the other hand, there will be no reduction potential of cost like that for the classical utility power conversion technologies, where larger power plants provided better economy. For PV technology on a small scale, new ways of cost reduction must be found in the long run.

The fact is, however, that during the last two decades of PV market development (with reasonable growth rates, as we have seen), the BOS costs have stayed around 50 % of the total system cost. So BOS has kept pace with overall development, despite the fact that the module costs have been reduced by almost one order of magnitude over that period.

That this parallel development can be kept over the next order of magnitude (facing another two-decade period) of the total cost, is often anticipated but far from certain. Here obstacles as well as chances must be looked for in closer detail.

In principle, in the system approach, components which allow higher standardization for mass production must be separated from systems where BOS parts with highly individual adaptation are involved (this also implies mixed situations).

Standardization is highly desirable and may be more easily achieved for

- inverters
- string/DC parts
- infrastructure parts (roofs/facades, etc.)

while individual design is typically required for

- PV fed/integrated mobile appliances and equipment
- autonomous /remote power supply
- hybrid power supply systems
- special architectural integration

The first class of components immediately suggests a large cost reduction potential via mass production, which is derived from general industrial experience. Also for the grid-connected systems, a strong market pull is now guaranteed by the grid-connected PV programs with rapid development in technique and economy.

The second class, however, needs more subtle strategies of cost reduction, due to the higher technical complexity as well as to the individual access to smaller market volumes concerning a larger number of rather different applications. So, in general, progress in this area strongly depends on the individual economic conditions valid for each market segment.

It is expected that for the time being a stronger impact on mass production must be expected from the grid-connected development due to the several subsidies for PV generated electricity in the public grid.

A further principal parameter in the cost of a PV system are the auxiliary services and the non-hardware. As for any technology, higher hardware cost allows more expensive service. The manpower required for design decisions, planning, installation, commissioning, and maintenance is a relatively fixed rate for each system. Moreover, it tends to increasing specific cost (being, of course, rather different in each country). Their share in the total balance can only be reduced substantially through rationalized and more efficient organization and logistics.

One way leads to a high degree of standardization of the hardware itself, which in turn allows a simple design solution and easy procurement procedures for the private customer. This is the way for additional PV systems to be mounted on existing infrastructure (roofs, etc.).

However, one should be aware of the challenge hidden in this task which is often underestimated!

For example, if we take a 3 kW_p-system to be installed on a one-family building and assume an investment to be taken of around 1 Euro/W_p, which means an order volume of about 3 000 Euro. Comparing this with everyday selling, it would only allow a rather limited service and marketing effort for a distributor or supplier. Assuming a service margin at 10 % of the investment, only service worth a few working hours in total could be afforded for the whole deal, including planning and after-sales services (including, by the way, also warranty for this service!).

So, for the smaller grid-connected systems, other and more efficient strategies must be found in the long-term, and present-day PV business does not reflect future demand.

In the future systems must allow final selling from the warehouse with a minimum of service requiring some practice of the consumer to make its own installation and operation. Such a "plug and play" approach is well within the scope of PV. In principle, the grid-connected systems could be designed in such a standard manner to allow this.

For the total cost balance of a PV generator the inverter is of strategic importance, if modules become cheaper. The first development of PV inverters during the late 1980s and the early 1990s (where the so-called 1 000-roofs program has been launched in Germany) the new technology took much benefit from the technology of the small size power electronics for driving electric motors of variable speed, which were already on the market in large numbers. Following [Meinhardt 2000] the drive inverters are specifically cheaper by a factor of two to three compared to PV inverters. Of course, the latter have to cope with more sophisticated technical specifications (MPP tracking, high efficiency conversion for energy purposes etc.) leading to more expensive equipment.

After the first prototyping, PV inverters quickly became cheaper and at the beginning of the German 100 000 roofs program, the technology was ready to go for that market now paying off for independent development efforts.

From [Meinhardt 2000] figures for the historical development of specific prices of inverters during the period 1990 until 2000 are quoted (Fig. 3.5).

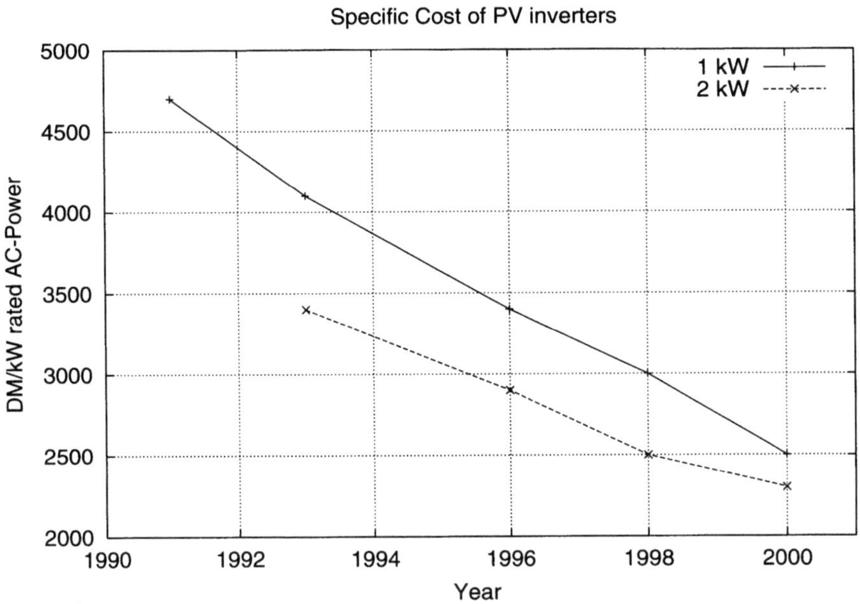

Fig. 3.5. History of specific cost of PV inverters in the German market (after [Meinhardt 2000])

The cost per kW_p (nominal DC-power) came from 3 700 DM down to about 2 000 DM (each for a 1 kW-system).

The question is now how these costs can be further reduced to match another cost reduction of at least a factor of two during the next decade. The following statements are quoted again from [Meinhardt 2000]:

The strategies for cheaper power electronics have to take into account:

- increase in mass production
- optimization of generator concept
- optimization of production (integration)
- innovative inverter concepts
- improved operational function and comfort
- increase of inverter size

The last requirement seems to be of interest, because it appears as a counterargument against further modularization.

Indeed, it remains a fundamental question whether the kind of technology related to power electronics follows similar scaling rules as the PV modules obviously do.

There are two opposite arguments involved in any "modular" concept:

1. the reduction via mass production favoring smaller units to the benefit of larger numbers in series production for a given production volume (e.g. expressed in MW)
2. the reduction via larger size units (related to the inverter concept and the specific cost of electronic components involved) following more classical rules of cost trends in power electronic

The matter is undecided yet leading to mixed concepts in practice. During the last years, a number of concepts and designs have been pursued, providing practical and theoretical experience.

For instance, for larger generators (100 kW and beyond), a so called "string inverter" concept has been introduced, separating the two functions of MPP tracking (via individual DC-DC components for each string of a generator, which are simpler and manufactured in larger numbers) from the power inverter feeding the grid and collecting the string output on the DC side. A good overview for a company related strategy can be quoted from SMA-Regelsysteme, a leading manufacturer of PV-inverters providing a cost target of 0,35 €/W_p until 2005 (Fig. 3.6).

In any case, these results of cost development and design have to be waited for.

Independent of the technical questions arising with PV generator concepts, there are also a number of new conditions of non-technical character influencing the cost balance substantially.

A certain barrier to PV installations by a technically skilled non-professional person are the legal conditions for a grid-connection and the necessity of monetary contracts. This never has been the case at the consumer level concerning any appliances in the household (apart from rather special equipment) up to a few kW. In the case of PV, however, as power is *generated instead of consumed*, the situation is different. The installation of a generating capacity usually is ruled by much stronger technical specifications (especially safety related) and also by a legal utility-oriented framework.

At present, the question remains open how these boundary conditions might change with the rapid development of the PV market together with a parallel historical change of paradigms in a liberalized power market opening new ways to decentralized electricity production in a quasi "fractional" power supply scenario. In such scenarios, however, the role of small power producers has to be newly defined technically as well as legally.

An example of such developments is given with the German 100 000-roof program, which required and triggered a number of legislative procedures to find appropriate conditions for the new market.

A consequence of this development is also that a completely new consumer "feeling" concerning electricity arises and has to be incorporated into the way electricity is sold and socially dealt with.

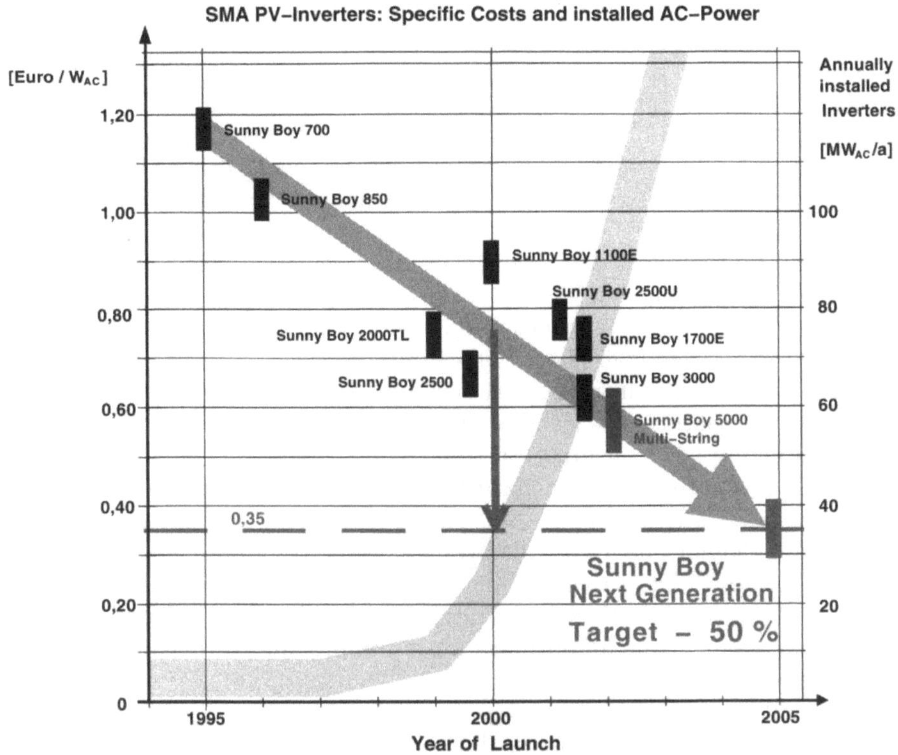

Fig. 3.6. Concept and Cost-Strategy of PV-investers (courtesy SMA, 2002)

These questions have immediate economic consequences, as they influence consumer behavior and consumer decisions as an investor, considering a typical private investment of the order of a few thousand Euro (a level of investment where in comparable cases no such regulations are usual). Even the consciousness that there is a role as an investor regarding power supply now includes generation at the consumer level for the first time.

An interesting question arises now from the fact that the legal conditions and the way contracts between grid operator and "consumer" are fixed strongly influence this role. For instance, if legislation promotes grid integration by subsidizing the price paid for each PV generated kWh fed into the grid, the owner of the PV system will optimize his economic yield. He will favor grid-counted kWh instead of optimizing his own electricity consumption, including his own PV power.

The present 100 000-roof program clearly shows this trend. Unfortunately the rather important possibilities to optimize and adjust electricity residential consumption comprising the whole household tend to be neglected in this way.

If, however, PV-generated electricity can be generated at cost comparable to the price level at which the public grid operator offers the kWh and no subsidy is given, then the situation is reversed again. So, in the scenario where PV becomes a cheap power source (favoring socalled "net metering" concepts), building optimization will automatically take place.

The introduction of such a new organization of power supply, however, is not merely a question of the technical options, but also a principal question of redesign and long-term optimization of the whole supply system. From a grid-oriented point of view, the decentral integration or PV generator capacity not only leads in practice to a different consumer demand pattern with reduced overall power demand but also to a structurally different load curve. For a grid and supply manager, it means that with a high share of PV capacity in the grid, the average load curve will change.

At the same time, if each consumer would optimize his load behavior to maximize the share of PV generated power he produces in his own system, then completely new consumer patterns will emerge.

Often, this scenario is criticized for focusing on the stochastic availability of PV power. This is generally true, but neglects that there is a strong stochastic behavior already in the present system to be managed by the utility, with quite sophisticated technical means to guarantee grid stability. It might even be possible that in an intelligent building power supply system, the grid demand might be optimized, even improving the capacity credit of the whole system.

A complete picture of all these possibilities has not been worked out up to now, but the questions remain on the agenda of a future public power supply scenario base on a much larger share of renewable energy ressources.

A much broader view is to be envisaged concerning future supply structures integrating demand and local opportunities of renewable energy supply. For the electric power supply central and decentral elements shall be integrated in a true modular approach [Kleinkauf 2000] leading to even higher grid stability and power availability together with a modular standard of equipment, which in turn allows much lower production cost for larger series. Here PV merges with other options like fuel cells or smaller district heating with combined supply of heat and power.

With respect to such visionary concepts of infrastructure, the innumerable possibilities, of how the PV hardware can be integrated into other hardware installed at the consumer level will generate new market segments for PV, and appropriately adjusted technical solutions.

An example of this is PV covered shingles designed according to typical roof architectures. This way has the advantage that even rather traditional residential building patterns and architectural esthetics can be satisfied in this way.

While a new PV generator today may confer pride of ownership, in the end PV will become more or less "invisible" as an everyday integrated and ubiquitous property of our infrastructure surface that is not paid special attention any more.

For these reasons, no simplifying rules should be fixed concerning strategic road maps to mass production of BOS components.

Inevitably all parts or components of a PV system which are contracted from the market of general technical equipment, e.g. metal parts, cables or general electric equipment for the assembly, take benefit from already existing mass production. This means that there is not left much potential of further cost reduction from innovation, but mainly by larger orders. For this reason, with shrinking cost share of the indigenous PV parts (modules, inverters), the economic pressure on their cost share will increase, especially for small PV generators.

Together with the anticipated market growth, alternative solutions which allow the substitution of such conventional components will become very interesting from the economic point of view. This trend holds, provided their potential in cost reduction is higher. We may expect, that in coming years, specific solutions integrating a whole bunch of innovative components will emerge in the different sectors of the PV market.

One of the most interesting markets will be the integration into buildings in mainly two areas. The first market segment is the typical one or two-family housing with a rather standardized roof. There an evolution of integration into the manufacturing and assembling process will take place in coming years. Increasing offers from manufacturers of prefabricated roof and wall elements or even completely prefabricated houses indicate this change.

Different from this is the market of PV generators in commercial buildings from simple manufacturing shops, warehouses or larger buildings. Usually, the commercial customer, if he thinks of PV at all, is also ready to accept PV as an attractive part of the specific architectural style with which he wants to augment the public relations of his company. This is proven by a rapidly increasing number of impressive architectural solutions in industry and trade around the world. In many cases, the "value" of PV concerning this PR effect is considered much higher than the still weak economic profit (if not loss or disadvantage).

A side remark concerning solar thermal applications is required. A large share of the energy balance of buildings is the demand for thermal power supply or, more generally, the demand for air-conditioning (in cold as well as hot climate). Today, energy efficiency is still rather weak and has to be improved. There is a strong connection between thermal energy demand and electric power demand, from the customer's economic point of view, but also from the engineer's technical point of view.

New solutions are to be found for a reasonable combination of these two demand sectors, leading to rational and efficient use of primary energy. By already demonstrated techniques of construction, the resulting much lower heating (or cooling) demand also leads to new solutions for the electric supply. In this context, PV is only one of the aspects of innovative building technolgy. As electric power represents a principal and basic quality of modern life, the combination of thermal demand and supply patterns (solar or environmental energy sources) together with the unique PV option will create quite new concepts.

If we extend our view furthermore to other options like fuel cells and local storage of energy, a vision of a totally new landscape or architecture with a maximally decentralized energy supply system is developed [Kleinkauf 2000]. Such a structure, by the way, not only allows integration of PV and a decentralized use of renewable energies, but also improves the stability of supply based on an integrated network of individual and autonomous supply systems.

Architectural integration of PV therefore will become one of the challenges during the coming years and an important component of a future energy network.

Various boundary conditions are to be faced concerning the large variety of PV applications for consumer appliances mainly with autonomous systems. Most of the consumer products of the modern world depend on electric power supply. Small portable appliances build the environment of modern man connecting him to a communication and IT-related network permanently active. Electric batteries are a key component of this environment and also an important economic part of these systems.

As the power demand of mobile equipment normally cannot be reduced without limit (there exist lower bounds to the power demand depending on the physical limits of the application), the primary power supply becomes an important factor. This holds even if the specific power demand is reduced, due to the fact that the complexity of functions and its comfort are increased in the same way. The period of recharging the battery normally is defined by consumer behavior and related possibility to bring the battery, as apposed to the equipment, to a grid-connected station for recharging. The periodic dependence of the equipment on recharging constitutes a major disadvantage of such systems. Any technological means to reduce this dependence therefore would be highly desirable.

PV forms a principal alternative to grid-connected battery charging. It is, however, not the ideal alternative supply option due to inherent deficits, mainly the insecure availability of sunshine or daylight or the improper or inconvenient time interval required for exposure of the equipment to light.

There are, however, also visions of a solar active environment for PV based power supply completely integrated close to the application. For appliances directly connected to the consumer, the PV active surface can be integrated close to the body clothing or to any mobile or carried equipment with a surface exposed to the light, permanently or from time to time.

Many aspects or technological requirements are not developed yet and need great efforts to be developed for such a vision. However, if we think of the innumerous applications, the development of an appropriate technology seems worth this effort and the solution may define another revolutionary synergy of individual mobility, communication, and independence from infrastructure.

One of the key requirements can be directly derived from the experience with progress in electronics. Certain standards of components and building blocks must be developed to cover rather diverse applications. This holds for storage as well as for the primary power supply units. Furthermore, it might also be reasonable to integrate more mobile energy resources than only PV, leading to higher availability. In recent years a number of options have been discovered or reinvented to be useful as a power source.

In the end we may imagine a futuristic picture of a human body covered by a textile equipped with various integrated abilities related to comfort and communication or even intelligence requirements independent from any other supply of energy. The vision is not so far ahead, as designers already started to create such "clothes" recently.

Even if PV only represents a smaller part in this vision, it could serve as one of the technologies to realize this. The breakthrough of such applications will create another branch of future PV, leading to quite different technological evolution. The main criteria and boundary conditions will then be like area or power-related specific weight, flexible cell structures and others not imagined yet (space applications may give hints to such requirements). Thin-film technologies might find a unique market of application in this area.

This qualitative evolution of PV in quite separate markets surely will enrich the general progress. Finally we should be aware that the technology of PV is still in the beginning, and will lead to innovations not imaginable from the present representations.

To complete the system approach, we finally turn once more to the big scale.

Very large PV generators have not been attended so much in recent years according to the general conviction that better economic chances are given for smaller generators. This view is in contrast to early discussions about the prospects of PV concerning the large potential also at the utility level in the multi-MW range.

Furthermore, a very large generator at a given site is assumed to be exposed to larger stochastic variations in climate and weather, reducing the relative capacity factor and leaving PV the minor role as a "fuel saver." This is a fate shared with large solar thermal power plants, for which no real access to the electricity market has been opened despite the fact that their actual economy is much better than for photovoltaic.

However, one should not underestimate the long-term potential also for large photovoltaic generators after longer progress along the learning curve.

In principle, for very large mass production of PV, the cost for large generators will also provide advantages for the use of wasteland in sun-rich areas. The economic advantages of larger PV utilities have to be verified and the large generator concepts pursued in due time.

Concepts including the use of free land for power production and general "energy farming" should also incorporate PV as an option from the beginning. In the future, where PV technologies with much higher efficiencies (also taking concentrating and tracking concepts into account) are available together with large options of storage (hydrogen, etc.), PV will play a role.

It remains a specific advantage of PV that the basic building blocks, the modules, are now open to mass production in a real market, steadily improving the quality and reducing cost. This inevitably will also lead to large scale applicability and competitiveness.

System technology will meet the challenge of cost reduction in parallel to the reduction expected for the modules by technical innovation and mass production already visible ahead. This will hold at least for the next stage until 2010. The industrial structure, however, strongly favors larger enterprises for component manufacture.

Beyond 2010 with PV markets in the multi-GW range, more sophisticated production techniques for BOS components have to be followed, especially for smaller generators in the kW range and below, where non-hardware related cost remains a problem. To reduce necessary services in planning, installation, maintenance, etc., quite innovative marketing and selling methods have to be realized. These problems must be solved together with corresponding technical solutions (largely standardized and modular system concepts). PV must become a natural property of general infrastructure not paid much attention but being an element of quality (in the building sector, in communication systems, etc.)

New and creative niche markets for widespread PV applications have to be opened, where PV provides unique advantages, especially in decentralized applications and for autonomous systems. In these cases, action can be expected earlier due to the fact that cost sharing in autonomous systems is much in favor of PV. This holds already for an increasing number of PV applications (from mobile phone transmitters over signal units to PV powered parking meters, etc.). With each cost reduction in module production, "primary energy" photovoltaics become cheaper and offer easier access to such innovative solutions.

Last but not least, photovoltaic thin-film technologies will provide quite new perspectives with a much more flexible technical specification (concerning electrical design) as well as with freedom in module size together with flexible geometry. If, finally, PV modules with flexible substrates become available, new applications will be accessible and old ones revolutionized.

3.4 The Export Drama of Photovoltaics

Before coming to final conclusions about the PV industry and the future cost reduction potential, some remarks are devoted to the history of PV exports.

From the beginning PV has been considered a technology especially suitable for sun-rich countries and decentralized off-grid applications. There is no doubt that PV provides power supply for such occasions, and a number of niche markets have indeed been developed.

That was also an argument from the beginning that PV should exclusively be devoted to such applications, supposing that PV can only become economical (which means competitive) under such circumstances.

As a matter of fact, these conditions are largely given in developing countries where a majority of people live "off-grid" and with almost no chance of getting "on-grid." This suggests that there is plenty of market for PV in the developing countries under "favorable" conditions.

Even if the argument is true from a technical point of view, it becomes also somewhat cynical from a social point of view because the local "markets" of energy demand are not favorable to a technical solution demanding long-term investment due to a chronic lack of capital.

Meanwhile this lack of proper instruments of capitalization of PV in the developing countries has been paid more attention, e.g., in programs of the World Bank and other financial institutions in recent years. So there is hope now (proven by impressive examples of successful distribution of solar home systems in an increasing number of countries) that PV will overcome the historical delay in opening one of the "indigenous" markets.

The early starts of PV in the developing countries are largely quite depressing stories. Many of the technological engaged companies tried to enter this business and a similar number have also been retreating again. The reasons of disappointment were manifold and of technical (underestimating harder conditions of installation, operation and service) as well as of non-technical nature (lack of educated operators, of long-term orientation of investment, expensive services, etc.).

However, in the long run, these difficult markets will be developed and must be paid attention by an early stage of technical evolution to have appropriate systems available in due time.

All companies are in watching position for the difficult export markets and also in a competition as global players.

What are now the strategic conclusions with respect to export?

This question brings us back to the typical structure of a photovoltaic system which is relevant over an extreme bandwidth of system sizes holding for the small solar home system as for a large grid-connected PV generator.

There are still

- PV modules
- inverters

- batteries
- control electronics

to be implemented in any PV generator.

For any export strategy it must be decided which parts of the system will be manufactured locally and which will be supplied by import to the country. There it has to be taken into account that the relatively high investment of a PV system makes a higher share of added value within the domestic market more attractive. On the other hand, in many cases, the enforced domestic production of BOS components generates exhausting problems with quality and total system performance, often spoiling the success of market introduction.

There is, however, no alternative to the development of solar markets in developing countries. Technically it does not represent a true difficulty any longer. It is a question of political and industrial engagement to evolve the strategy. The more than two billion people in the world without access to even the lowest level of power supply cannot be supplied by classical means. Instead, decentralized power supply must become an essential part of a development strategy. And decentralized in this case also means the extensive use of local renewable resources, solar radiation being the largest and ubiquitous one, together with a real chance of local production of systems or at least important components.

Furthermore, the modular character of the PV system provides a unique chance of harmonization of global market forces with development tasks, because no other technology can fit both demands in one technology produced at industrial scale.

The integration into systems can take place at the local level, providing also a chance to implement a larger share of local added value in this process. It is to be noted that the typical comparative price level of electricity for the smallest consumers far off a public grid is already much higher (e.g., diesel-powered batteries, etc.) than for grid-connected supplies. It is not a question of subsidizing the power but establishing a local market with appropriate financing of new techniques.

Practically all globally active PV companies are engaged in developing countries, and are aware of the above mentioned difficulties but also of the market promises. The battle at present seems to be more concentrated on the technological competition in the industrialized countries capable of the high-tech approach for the module technology.

The c-Si based solar cell technology fortunately allowed an easy sharing of manufacturing between industrial primary cell production and the final module assembly in a local factory in the developing country. This sharing has been praxis in most of the industrial PV engagements in developing countries so far.

However, in the case of thin-film technologies, the fully integrated production of complete PV modules does not allow this separation in stepwise production. Instead, any thin-film PV manufacturer must decide, where production will be established. As long as production is still on the pilot level (which according to the dynamic development will remain the case also for coming years) the risk of transfer of technology to another country will be high preventing an early engagement or at least restricting it to the bigger emergent nations (as is happening for India, China, etc.).

From a technical point of view, the transfer of production in the long run will be dominated by economical structures. It is to be expected that for thin-film technologies, the transport and distribution of bulk material in PV modules like glass substrates over longer distances will be uneconomical. In this case, the deployment of production capacity in a country where such bulk materials are locally available with suitable quality will become more economical than the transport of complete modules (if market volume is large enough).

It is to be noted that this argument is limiting up-scaling of manufacturing capacities. In [Bruton 1996] the comparison of up-scaling demanded by a cost level of around 1 EURO/W_p of module production lead to the scale of 500 MW production capacity for the c-Si technology. From the export point of view, it appears unrealistic to establish such a capacity within one manufacturing unit to supply the major part of the world PV market from one place on the planet, and a market volume justifying such capacities is not immediatly ahead.

The cheaper PV becomes, the more unlikely will be the long-distance transport of PV modules, and a splitting of manufacturing capacities will take place. Where the optimum of distributed manufacturing capacity will be set remains open.

This conclusion is also supported by another aspect of industrial learning. The same argument that leads to learning via mass production of the modular product itself also holds for the production equipment itself. Learning means that the design and engineering of production equipment will be also changed, leaving the pilot stage and approaching series production.

The question of PV production cost will be in this case

> What is the investment for a second, or third production unit of equal design?

At present, practically no system part of the production units for thin-film technologies has been manufactured a second time. This is different for the crystalline silicon technologies, where the production equipment after twenty years of commercialization is now manufactured in some numbers (e.g. crystal growth equipment, solar cell furnaces, module laminators, etc.) for an increasing number of industrial customers worldwide.

In principle, when leaving the prototype or pilot status, the PV industry also induces a primary market of production equipment, which is of high interest to industrial production itself. This holds especially for the industrialized economies leading technology development.

Therefore, export does not only mean export of solar products like PV modules, but also export of production equipment. This can happen as easily in competing countries as in purely importing countries. The competition in this area is part of the global race for technological leadership between firms and national economies.

In the end, for countries like Germany, the export of production equipment is a typical final stage of international "work sharing" leading to highest benefits for the global partners in this "game." Apart from exporting PV products, which is dominating the first stage of PV development, a complementary strategy of successive transformation of exports to more and more production equipment will be inevitable (a process which can be observed for instance in car industries, where car manufacturing as a technology is steadily transferred to the emerging industries in developing countries while the production equipment itself is imported).

Adding up all the arguments concerning export, we come to the following conclusions:

> The industrial structures have to be established in the industrialized countries first, where for research-intensive manufacturing the best conditions are available. No manufacturer can bear the risk of going into export with a technology not mature yet. Instead, export can only become a success if the technology has been fully qualified under favorable domestic conditions with full backup by research and market implementation programs.
>
> On the other hand, export strategies have to be started early by establishing cooperation in due time and starting business with well adapted local industrial partners. For this reason, a good synergy has to be found on the political level between the national programs supporting the industrial development and the export-oriented supporting measures.
>
> The enterprise structures required for export strongly depend on the quality of local production. Any investor aiming at import of PV into emerging or developing countries has to decide what cost and technology sharing reasonably can be achieved and how the partner in a specific country of interest can be qualified.

No PV industry can afford to neglect the export markets in the long run. Due to the now given better conditions in a number of industrial countries subsidizing PV, there is a temptation to neglect exports. This especially holds in Germany, where no national strategy for the export of solar technologies even into developing countries exists. This is in contrast with the national programs in Japan or US, where clear commitments are implied in the general strategy aiming at export and the developing countries.

3.5 Strategies for Photovoltaics Industrialization

Let us again cast a brief glance toward the specific story of PV and briefly summarize the first promising decades of industrial development:

1. Military application produced the first big push to production and technology (taking the rapid development of satellites predominantly as a result of the Cold War).
2. The PV based power supply of satellites even overrun the nuclear option with inherent advantages.
3. Subsequent "terrestrial" spin-off was guaranteed for a number of small niche markets, where PV could play its advantages and compete with conventional techniques at rather high cost of electricity (typically applications far off the grid and at low rated power).
4. The first oil crisis in 1973 brought the proper "political" arguments for a far-reaching strategic goal, namely independence from limited fossil resources together with the dawning of an environmental protective view, both later merging in a holistic approach called "sustainability."
5. In the last two decades of the past century, PV was placed in a number of niche markets securing a steady growth of production (at modest rates) following the learning curve.
6. Political programs for grid-connected PV started in the 1990s, in US, Japan, and Europe, leading to substantial growth, dominated even by the least economic application in public grids.
7. During the 1990s new technologies also came up to the level of industrial pilot production, the most important ones being thin-film solar cells.

There were also disappointing experiences:

1. The market considered the most natural for PV, off-grid electrification in the developing countries, did not really take off, after many attempts.
2. The technical evolution of solar cells took a slower course than anticipated by research results and by the knowhow already achieved in the 1970s and 1980s (remember the high expectations in the early 1980s, e.g. for ribbon-grown silicon or amorphous silicon thin-film cells).
3. The large majority of utilities in the world still consider PV an exotic "toy" not to be taken seriously as part of a power supply strategy on a large scale.
4. PV as a principal long-term investment has to face alternative opportunities for investors during times when energy is relatively cheap.

For many people, the military aspect is considered a tragic one. However justified this may be, the military approach has been rather important for PV (as for many other examples, war might be the "father of all things," following the ancient Greek philosopher), leaving only one simple decision far below the level of complexity quoted before:

> If it is technically feasible in due time to provide a profitable military advantage, it will be done at any cost.

The military budget for PV development in the 1950s and 1960s can only be estimated, and is perhaps of the order of a billion dollars. The result, however, led to a proven technology, mono crystalline silicon solar cells and modules, and also the start of a manufacturing industry (literally!).

It may be speculated that PV probably never would have been able to beat the high cost threshold by starting immediately "on earth" and that the arguments might be the same as claiming that PV had no chance to become feasible at reasonable efficiency and economy.

The fact that PV was a proven and available satellite technology was decisive for the subsequent "terrestrialization." At the starting point, the specific costs of generated electricity were about two orders of magnitude higher than bulk electricity from fossil power plants still hampering a serious political discussion about PV as a future global option.

Only when the first oil crisis happened to have a grip on the industrial world, the alternatives began to be discussed in a broader context. This at least triggered a first non-military approach to PV technology, with excellent scientific and technical results and a number of demonstrative applications. Looking back, it can be stated that the process of development of PV has become quasi-irreversible, despite uncertainty about its future pace.

An important question about the historical process is whether there had been a real possibility to accelerate it. In the context of the US programs during the Carter administration (highlighted in the important "Global 2000" report) it may be questioned, wheather the scenarios growing from the goals of the program would have been realistic if pursued in due time. One should recollect that there had been scenarios claiming installed capacities for PV in the multi-GW range until 2000 leading to competitive electricity cost via mass production.

Neither did the US administration execute this original PV program until 2000 (instead the money was spend for Star Wars), nor have the PV cost scenarios been realized.

What can we learn or derive from this history?

The appropriate strategy for a pathway leading to higher mass production of PV and thereby to lower production cost on an industrial scale includes several elements concerning the complexity of the "learning" process.

The key factors discussed in the preceding chapters were:

- market development
- development of applications
- quality of the learning curve
- speed of technological innovation
- industrial investment cycles

There are more factors of influence which cannot be predicted but depend on more political strategies and movements, i.e.

- consumer behavior
- influence of political programs
- policy of utilities
- national political boundary conditions

The basic obstacle to further development of photovoltaics may be summarized with the so-called "Hen and Egg Problem," which means that the introduction of a new technology is hindered by the high production cost at low number of pieces leading to non-competitive high prices in the beginning. In general, whenever a new product based on a new technology with high initial development cost has to compete with commonly introduced and standardized products, the introduction will face a big problem. Whatever the strategy will be, the early investor has to bear the risk of a large pre-investment in development and marketing over a certain period of time often not assessable in the beginning.

In the case where a new product is competing with existing ones without at least one specific new and advantageous property, the new technology faces the highest barrier. The new product has to be subsidized over a long period, until by establishing adequate mass production, it reaches the point where it overruns the existing old product with lower cost and higher mass production. The reason for the introduction of a new technology, if it is more expensive without promise of near-term cost reduction, is normally based on political arguments. Over long periods of industrial development and for many key technologies today fully implemented, the most common and typical arguments for starting them came from military prospects. Only recently new arguments in the energy sector stem from environmental considerations and from long-term risk evaluation.

If a new technology, however, provides special features and advantages compared to the old technology that are obvious to the consumer, then the startup period may be shorter and the investment risk lower (also requiring fewer subsidies in case of an accelerated strategy for political reasons).

If we take the political premises first, it is obvious that there is an indigenous national interest to keep knowhow and industrial capacities in the country, aiming with high priority at "new markets and employment," hopefully combined with a (more or less nebulous) environmental policy. The latter policy, even if stated officially, does not produce clear boundary conditions for technology development. Such a national point of view is a major driving force for a global technology competition and is clearly also a point in the case of photovoltaics.

Another example of political boundary conditions is the debate over the question of where and how to start a suitable strategy for CO_2 reduction. It is a global problem, and many argue that the reduction of CO_2 emission must be ruled by economic evaluation of all available technical options. In such cases (a policy now called "clean development mechanism"), of course, preferably technologies already available and cheaper today will become winners in this game. Just emerging technologies which have to be evaluated by their general technical potential or by their development chances probably will not be taken into account by the investor.

Furthermore, any new technology which is more expensive or demands a long term introduction strategy requires more resources from the national gross product on a shorter time scale. Such a technology therefore also loses ground in the political evaluation process which finally sets priorities for the public budgets.

Consequently, and repeatedly, it is argued that PV, which today is still the most expensive way to generate electricity, should be the last option to substitute for fossil power now.

On the other hand, it is also clear that many of the technologies which are preferred following this line of reasoning do not solve the CO_2 problem in the long run. At best, they help to delay the global effects and thereby help to postpone a thorough discussion about the effects. For an integrated evaluation of alternative options, one always has to define the period of integration. Therefore, any discussion on photovoltaic technologies in the context of the energy supply balance requires a political time horizon of at least 20–30 years (as analyzed in Chap. 1).

A thorough discussion of the relevance of PV to the long-term solution of a problem like CO_2 reduction requires a well-based scenario analysis about the technological and economical potential of all technologies in the long-term and to assess their uncertainties.

In all aspects there are "hidden" typical time scales of decision, investment, political emphasis, and last but not least, scientific progress. Together with a given historical scenario, these factors are interwoven into the "multidimensional chicken and egg problem," not easily to be solved or transformed into a reasonable strategy.

Furthermore, the development of such a strategy depends on a concrete situation of communication between different "players" in this game (some of them undergoing a difficult and risky process of globalization). It is interesting to observe that for instance in different but comparable industries or countries, the strategic conclusions and measures are sometimes quite different from each other.

It is therefore an important conclusion that there is no simple strategic rule based on a clean intelligent decision. Far more, it will be a complex process to start a technological progress and how to provide proper conditions for a continuous success.

After these general considerations and warnings, we now come back to the case of PV. There are promises and obstacles as well.

An important argument throughout this chapter has been the analysis of the "learning curve" for photovoltaics compared with the general features of such learning processes in other technologies.

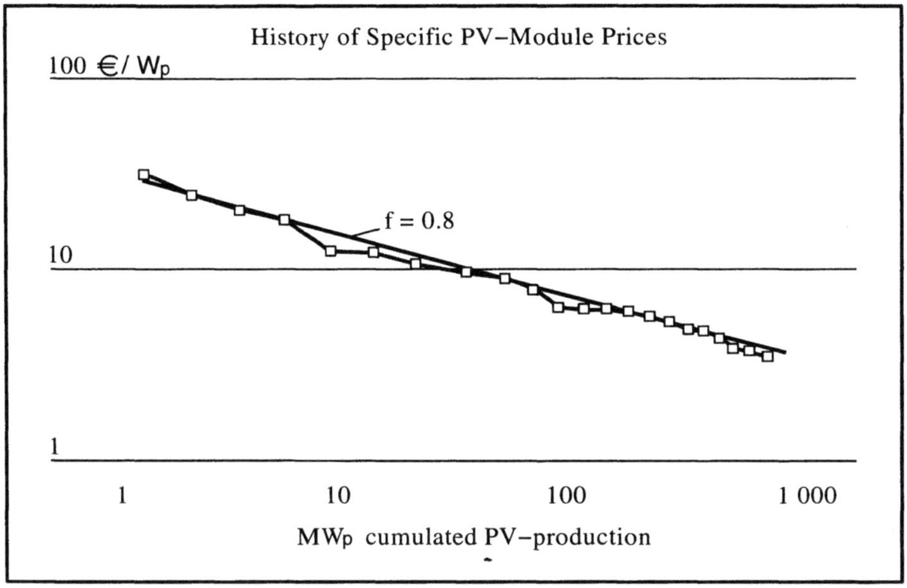

Fig. 3.7. History of PV module price learning curve (by courtesy Bayer [ASE 2000])

It is quite suggestive that the learning curve describes a fixed track which PV development has to follow, no matter how fast. The speed of development could be interpreted as a pure consequence of the amount of public and private financial push. For the last two decades at least, the history of PV confirms this (Fig. 3.7).

There remain doubts that this argument is truly independent of the circumstances. Despite the fact that the historical track of many technologies seems to confirm the pattern of learning curves as a universal one.

Looking back, the learning process always confirms a historical path of this type. But for logarithmic scales one must be careful about statistics and small deviations in the slope of learning curves may lead to substantial differences in the cost projection over decades.

The learning curve implicitly assumes that technological development keeps pace just in time with the market, thereby guaranteeing steady progress for production and for productivity. In reality, technological progress always becomes a mixture of increasing productivity by up-scaling together with a continuous change in technology, both parts following their own rules and depending on various boundary conditions.

For the scientific and technological approach, on the other hand, the time scale is set by the rules of R&D organizations, determined by other qualities like excellent PhD work, a well-equipped laboratory, a fruitful academic team spirit, investor's personal spirit, and so on.

The scientific world follows different rules, for instance the ambition to be the best or free information exchange favoring early publications.

And last but not least, inventions and innovations are not calculable and do not follow continuous and predictable courses.

In contrast to this, evolutionary progress in industry is much more governed by a reaction to the market and to competitors maximizing profits. Innovation is triggered by the chance for a higher or faster profit, prohibiting free information exchange and general progress. Industrial evolution follows evolutionary principles and driving forces different from the scientific approach. Taking these differences into account, it is by no means simple to organize and support the transfer of knowhow between these areas and to find proper rules for this game.

It was a specific experience for the PV industry that the technology first had to be brought to the possible customer. It was not yet the market pull which accelerated the introduction of PV or at least it was not the market responsible for the speed of progress (as is often assumed in the case of electronics, where things are also not as simple as they seem). It must be stated again:

1. PV cannot be simply let to the market forces, as the newcomer has to compete fully with the existing and largely introduced technologies producing goods of high standard, the kWh of bulk electricity
2. Instead, if there are (whatever) other reasons to implement and develop PV and if waiting makes things worse concerning such arguments, then learning must be accelerated beyond the market.
3. If acceleration is supposed then the acceleration program has to be politically implemented and optimized.

In this context, in the industrial countries, the earlier debates over the long-term chances of PV brought up various programmatic views, which may be classified in the following way:

1. a broad technology oriented public funding to accelerate the creation of proven techniques within existing industries (the approach, for instance typically followed in the US).
2. a national strategy connecting research, industry, and market introduction in a closed strategic concept guided by government and big national technological companies (could perhaps Japan).
3. a more scientific approach, assuming that the technologies available so far do not provide the proper basis for large industrial investments (near to German policies).
4. PV considered a matter for utilities, who decide about suitable investments in the energy sector (partly Germany, also US).

All of these arguments tend to the effect that the real long-term potential of PV tends to be underestimated in comparison with other options, and have been responsible for the relatively slow progress during the first decades.

That process can take place quite differently can be seen from the example of wind power, which took a surprising jump over only ten years, starting at a neglected level. It is also interesting to observe this progress in other countries, with different boundary conditions.

It was the US that started in the 1980s with tax credits to encourage private investment on a larger scale for wind power. Installed wind power grew rapidly to the GW scale, but on a weak technological basis, leading to great economic problems.

The early US boom in wind power, however, encouraged countries like Denmark (an early starter with an old tradition in wind power) to export machinery, accelerating the "national learning curve." Germany came in late but with a rather efficient national program (a technology-oriented subsidy followed up by the public feeding law) overrunning all utility complaints. The more than 2 % wind share in the public power balance and the roughly 2 billion DM turnover of the German wind industry makes it one of the unique success stories in renewable energy technology.

The story has been confirmed by the way that under proper technological and political boundary conditions, a real transformation in the energy supply system can be achieved within a few decades. The success of wind power has also convinced a growing number of utilities to change their policy concerning renewable technologies.

On the other hand, the example of wind power should not naively be extrapolated beyond the specific historical boundary conditions and context. It must be kept in mind that small series production of (small) wind energy converters a decade ago was still based on quite specific technological and industrial conditions:

- continuous research effort over many years, providing the proper engineering skills for the industrial start
- the fact that important components of wind converters could be taken from the stock of industrial production experience and similar theoretical ground (aerodynamics, structural mechanics, power electronics, etc.)
- the level of generation cost of wind power missed the economy by a factor of two compared to PV (still a factor of ten missing).

These conditions allowed a start for wind power engineering with basic design rules to be followed, and a political time scale of "learning" investment of only ten years.

In contrast to wind power, PV represents a completely new technology of power conversion and supply, with completeley new "rules" of engineering. It is this basic character of innovation which makes it difficult for many engineers, managers, and politicians to accept the change and to recognize the specific advantages.

What should be learned politically from the case of wind power is the following:

- Policy can decide upon boundary conditions for something new in an appropriate way. In the case of PV the new strategy has successfully been entered but a long-term concept is not yet in hand.
- The industry has taken part and is strongly engaged in the strategy. However, the industrial process must be seen on a long-term basis, which naturally cannot be the primary concern of the industry, wich depends on much shorter terms for profit and market.
- Therefore the present phase of industrial engagement must be politically supported (not simply by subsidies but more importantly by a long-term political commitment, which by the way is not unusual in questions of energy supply, even more when connected with environmental and security questions. Energy remains a key factor in industrialized societies, that has always been a political matter.

Perhaps a suitable PV strategy should be oriented on the example of building specifications now set by German law for heating systems. Despite the fact that energy is still relativly cheap, legislation imposed strong requirements on thermal standards for buildings and on the efficiency (together with standards of emission) of heating equipment. These standards were far beyond actual economy (for that, a much higher energy price level had to be assumed) and in reality meant an additional financial burden for the investor.

The measures in these thermal standards were only justified by a general strategy for reduction of the dependency of the national economy on primary energy imports as a national goal triggered by the first and second oil crisis in the 1970s and early 1980s.

In this context, it was clear that change could be only achieved by a long-term strategy concerning the life-cycle periods in the building sector (more than fifty years). It also had to be started at a time when energy is still cheap enough that the measures would be achieved by market forces in due time.

The policy should be seen as a success and the standards set as a "wise" conditioning of our energy system. Also from the point of "learning investment" the political investment of the national economy may be efficient, even if it was not economical in present terms, because in the long run, savings in energy cost, the political strengthening of our economy, and last but not least, the gain in industrial fitness with better technologies may well pay off.

Finally, the scenario for the period until 2010 is revisited again and the collected arguments may be summarized in the following statements:

Industrial management in the PV business is generally convinced that

- the market will grow at least at 20 % per year in coming years (fitting the scenario of Chap. 1)
- crystalline technologies will remain the workhorse of PV within this decade and will be able to halve the production cost of PV modules and systems
- thin-film technologies will arrive on the market within the next 5 years, and break even with higher shares of the market around 2010

- the learning curve continues to follow the historical path within the next decade; during this period, the PV market and growth is dominated by grid-connected programs in industrialized countries trying to establish domestic production capacities with global competitiveness.

Generally, higher figures may be realized under more favorable conditions. Any higher ranking in this competition demands a national policy that takes into account favorable conditions and strengthening of the R&D basis for the present industrialization as well as for the long-term innovations required.

For the horizon beyond the year 2010, the crucial question is whether this learning process can still be continued in the same manner, with further strong reduction of the specific cost and another market expansion into the multi-GW scenario.

Is there enough technology potential to achieve this? Where will the main markets for such growth be? And finally, does this process still fit into scenarios where there remain enough resources to support this development in the context of a general energy policy?

If one assumes the learning curve being expandable into the multi-GW "window," specific PV system cost of around 1 Euro per W_p are feasible, resulting from a complete diffusion of PV a large scale and in a large number of market segments.

In Central Europe such system costs would provide specific electricity costs around 0.10 Euro per kWh, and in more sunny climates in southern Europe even less thereby approaching the level of customers electricity consumption tariffs.

The expansion of the manufacturing capacities for PV modules does not set true limits. However, industrial managers usually hesitate to make forecasts for such long periods or up-scaling reaching beyond the investment horizon they deal with. Another source of uncertainty arises from the fact that with conventional silicon wafer technologies, a saturation level of innovation may be expected. Turning the argument around, it means that breakthroughs of new technologies are required.

Such breakthroughs are required as well for module technology as for system integration.

At present, it can be ascertained that the thin-film candidates of today will become major players in the period 2010 to 2020. However, it cannot be ascertained that the now chosen candidates will survive as the final winners. For further cost reductions beyond 2020 and at production levels with individual manufacturing capacities in the then-necessary range of a GW per year new basic innovations probably have to found.

Concerning the competition of technologies, it must be mentioned again that there will not be only one winning PV technology. Instead, different profiles will allow all technologies to survive in special market niches.

What seems clear is that such far-reaching technologies can no more rely on the present crystalline wafer-based technologies, due to elementary economy of material and energy balance. Instead, any long-term PV module technology will rely on thin-film technologies in one way or another.

From elementary physical considerations, the elements of further thin-film innovation must relay on

- thin surface film of active PV (few microns, if not nanometers)
- high efficiency (multilayer cells with different band gaps to cover the solar spectrum) approaching 20 %
- low temperature processes to minimize energy demand for production
- ample material sources to avoid shortage at large scale
- high cell stability to guarantee best life-cycle cost
- easy up- and down-scaling in module size
- flexibility concerning integration with alternative substrates

Additional systems, not to be neglected, may be low cost concentrating structures or multiple function surfaces (combining heating, electricity production, and shading, etc.) leading to total higher added value of any such surface utilization, or to better performance in large PV generators for utilities.

Whatever the candidates for long-term PV, the research to qualify them must start now. This is the lesson learned from the past 50 years of PV. For example, CIS technology started already in the late 1970s when a thorough screening of a large variety of chemical compounds was evaluated for their potential as photovoltaic devices. It should not be forgotten that also a large number of early trial and errors was necessary to qualify a handful of nowadays technologies.

The next conclusion is that even if a number of candidates have been successfully selected, the long-term race remains open in the sense that general scientific progress may open quite new options not imaginable today. This demands continuous basic research not limited by present "realism".

Another conclusion is that there is no easy way to a proven technology, whatever its promise. One of the reasons for that long "time constant" of PV development is that the product "module" is really a complex series of production steps from raw material input to module combining quite different technological sources, requiring high skills of adaptation from one step to the next. This is often a pure empirical learning curve without much help from earlier stages of experience along the learning curve. So up-scaling of this production must be performed carefully, avoiding jumps of hazardous risk for the sake of a steady controlled learning process.

The strategic path beyond 2010 has yet to be found in due time to guarantee this progress, especially meaning that laboratory research must be intensified in accordance with market growth.

Finally, one should also keep a vision that physics might bring new breakthroughs. We should not be fixed to the present physical understanding of the photovoltaic process and the present semiconductor technologies and materials. Further research on large thin-film devices might allow new design criteria for more efficient conversion of light quanta into electric current, or even into chemical energy, as nature demonstrates so perfectly. From that point of view, we should keep the technological window wide, open looking beyond the present.

In fact, PV represents a total different way to treat energy, with physical roots different from classical (thermodynamic) conversion, as well as with integration into our technical civilization. This interpretation is now entering the consciousness of energy management.

To make this vision come true, a comprehensive policy is required with a time perspective not stopping at 2010 or 2020 (representing the present stage of discussion defined by the time span engineers consider today), but setting goals in energy policy far beyond that horizon.

Such a policy has to include

- research capacities and abilities
- industrialization
- national energy policy
- export strategies

There is conviction that the appropriate cost scenarios and production goals can be realized. To reach a level of production cost even ten times lower than today, however, requires more effort with much broader scope of scientific and technological engagement. The level of present political engagement, despite impressive successes in the German market and of domestic industrialization, does not seem to be sufficient yet.

A real long-term commitment of the utilities for photovoltaic in Germany is urgently missing and necessarily has to involve a European perspective. A number of countries are within the process of definition of such policies, Germany being one of the most enterprising societies and best positioned economies. However, a well-defined German policy including all areas mentioned above is still missing.

References

[ASE 2000] T. Lauinger et. al.: “EFG-Silizium, Material, Technologie u. zukünftige Entwicklung”. Conference “Solare Impulse” Freiburg, June 2000.

[Beneking 1997] Claus Beneking: “Entwicklungs-Offensive Solar”. Studie im Auftrag des Landes Nordrhein-Westfalen, by AMO GmbH, Aachen und NEA GmbH, Herzogenrath, 1997.

[BINE 1998] “Kleine netzgekoppelte PV-Anlagen im Breitentest”. Infoservice BINE, FZK Karlsruhe 1998.

[BINE 2000] W. Knaupp, F. Staiß: “Photovoltaik – ein Leitfaden für Anwender”. 4. Auflage, BINE Informationsdienst, TÜV-Verlag, Köln, 2000.

[Bonnet 1998] Dieter Bonnet, Michael Harr: “Manufacturing of CdTe Solar Cells”. 2nd World Conf. on Photovoltaic Solar Energy Conversion Vienna, July 1998.

[Bossert 2000] R. H. Bossert: "Thin-film solar cells – technology evaluation and perspectives". ECN and Berenschot. Report to Novem, NL, 2000.

[Bradley 1997] "Renewable Energy – not cheap, not green". Policy Analysis No. 280, Institute for Energy Research, Houston, Aug. 1997.

[Bruton 1996] Tim Bruton et al.: "Multi-Megawatt Upscaling of Silicon and Thin-Film Solar Cell and Module Manufacturing – MUSIC FM". Final Report to the European Commission BP Solar (coordin.) Sunbury on Thames, UK, July 1996 (unpublished).

[Claeson 1999] Ulrika Claeson: "Using the experience curve to analyze the cost development of the combined cycle gas turbine". May 1999, s. [Wene 2000].

[DeMeo 1984] E. DeMeo, R. W. Taylor: "Solar Photovoltaic power Systems: An Electric Utility R&D Perspective". Science, Vol. 224, April 1984.

[Dimmler 2000] B. Dimmler, M. Powalla: "CIS-Dünnschichtmodule". Themenheft 2000, Forschungsverbund Sonnenenergie. Strategien zur Kostensenkung von Solarzellen. Berlin 2000.

[DOE 2000] "Million Solar Roofs Initiative Goals – Benchmarks". `www.eren.doe.gov./millionroofs/benchmark.html` Washington, July 2000.

[Dreesmann 2000] G. Dreesmann (ASE): "Vom Öl zur Photovoltaik". Fünftes Kasseler Symposium Energie SystemtechnikISET, Kassel, Nov. 2000.

[Francis 2000] Lisa Francis et al.: "Opportunities for Cost Reduction in Photovoltaic Modules". Arthur D. Little Inc., 16th European PV Solar Energy Conference, Glasgow, May 2000-10-17

[Fuhs 1998] W. Fuhs, R. Klenk: "Thin-Film Solar Cells – Overview". 2nd World Conf. on Photovoltaic Solar Energy Conversion. Vienna, July 1998

[IEA 2000] "Trends in Photovoltaic Applications". Report IEA-PVPS, Task 1, Paris, Sept. 2000.

[EU 1998] F. Staiß, J. Springer, H. W. Schock: "New developments in the field of photovoltaic cells". Working paper to the European Parliament, Directorate General for Research Energy and Research Series, 109. Luxembourg, Aug. 1999.

[Johnson 1995] Bob Johnson: "Report on PV-Manufacturer Shipments". Rep. PM 40, Strategies Unlimited (unpubl.). San Antonio, Cal., US, August 1995

[Kleinkauf 2000] W. Kleinkauf, B. Burger, G. Cramer, O. Haas, P. Strauß: "Stromversorgung mit erneuerbaren Energien – dezentrale Strukturen und modulare Systemtechnik". Conference "solare Impulse", Freiburg, June 2000.

[KPMG 1999] KPMG, Bureau voor Economische Argumentatie: "Solar Energy: from perennial promise to competitive alternative". Final report, on behalf of Greenpeace Nederland Hoofdtorp, August 1999.

[Kraus 1999] M. Kraus: "Stand der Dünnschicht-Herstellungstechnologien im Bereich Photovoltaik und Bewertung im Hinblick auf die industrielle Fertigungsreife". Fraunhofer Gesellschaft management GmbH, Munich Dec. 1999 (unpublished).

[Lawitzka 1999] "Use of Experience Curves within Germany's R&D Programme". May 1999, s. [Wene 2000].

[Martin 2001] Philippe de Renzy Martin, Interview: "We have a vision of growing to 100 MW in 2007". Photon International, Aachen, Febr. 2001.

[Maycock 1997] P. D. Maycock: "The Impact of Cost Reduction in PV Manufacturing and New thin-film Plants on Grid-Connected and Building-Integrated Markets". 14th European Photovoltaic Solar Energy Conference Barcelona, July 1997.

[Meinhardt 2000] M. Meinhardt, G. Cramer: "Kostenreduktion durch neue PV-Systemtechnik". Fünftes Kasseler Symposium Energie-Systemtechnik ISET, Kassel, Nov. 2000.

[NCPV 1999] National Center for Photovoltaics (NCPV): "PV Technology Roadmap". Workshop, Chicago, June 1999.

[Norman 1984] Colin Norman: "A cloudy Forecast for Solar Cells". Science, Vol. 226, p.319, Oct. 1984.

[Osborn 2000] Donald E. Osborn: "SMUD: a utility helping create the solar century". Renewable Energy World James and James, Sept.–Oct. 2000.

[Osborn 2000a] Donald E. Osborn: "PV as a core element of utility business plans". Renewable Energy World James and James, Nov.–Dec. 2000.

[Powalla 2000] Michael Powalla, Bernhard Dimmler: "Process Development of High Performance CIGS Modules for Mass Production". E-MRS 2000, Strasbourg, April 2000.

[PV-Produktion 2000] "PV-Boom beflügelt die Hersteller". Solarthemen 89, June 2000.

[PV-Market 2000] "PV-Markt wird sich vervielfachen". Solarthemen 892. Juni 2000.

[Rannels 2000] US Photovoltaic Perspective Workshop EU/USPV-Cooperation Anchorage, Oct. 2000.

[Ricaud 1994] Alain Ricaud: "Photovoltaic Commecial Modules: Which Product to what market?". 12th European PV Solar Energy Conference Amsterdam, April 1994

[Schmidt-Tiedemann 1997] J. Schmidt-Tiedemann: "Innovationen und Illusionen - Können neue Produkte unsere Standortprobleme lösen?". Physikalische Blätter 53, Nr. 12 (1997).

[Scheer 1993] Hermann Scheer: "Sonnenstrategie – Politik ohne Alternative". Piper, München, 1993.

[Schott 1998] T. Schott: "PV-Märkte und Industriestrukturen aus Sicht der Forschungsförderung". 13. Symposium Photovoltaische Solarenergie. Staffelstein, März 1998.

[Schott 2000] T. Schott, A. Bubenzer: "Killerargumente zur Fotovoltaik – eine Entgegnung". 15. Symposium Photovoltaische Solarenergie, Staffelstein, März 2000.

[Solarmodule 2000] Marktübersicht Solarmodule. Report in "Photon" magazine, 2-2000. Solar Verlag, Aachen, Germany, March 2000.

[Staiß 2000] F. Staiß: "Jahrbuch Erneuerbare Energien". Stiftung Energieforschung Baden-Württemberg (ed.). Edition Bieberstein, Radebeul, Nov. 2000.

[Wechselrichter 2000] Marktübersicht Wechselrichter. Report in "Photon" magazine, 3-2000. Solar Verlag, Aachen, Germany, May 2000.

[Wene 1999] C. O. Wene: "RD&D and Learning Investments: Findings from IEA Case Studies". May 1999, s. [Wene 2000].

[Wene 2000] C. O. Wene, A. Voß, T.Fried (ed.): "Experience Curves for policy Making-The Case of Energy Technologies" (ed.). IEA International Workshop, Stuttgart May 1999. IER - report, vol. 67, Stuttgart April 2000.

[Woditsch 2000] P. Woditsch: "Kostenreduktion in der Photovoltaik". Conference "Solar Impulse". Freiburg, June 2000.

[Woodcook 1997] J. M. Woodcook, H. Schade, M. Maurus, B. Dimmler, J. Springer, A. Richard: "A Study of the Upscaling of thin-film Solar Cell Manufacture Towards 500 MW_p per Annum". 14th European Photovoltaic Solar Energy Conference Barcelona, July 1997 (Proc. ed.).
[Zweibel 1999] Ken Zweibel: "Issues in thin-film PV manufacturing cost reduction". Solar Energy Materials & Solar Cells 59, 1999.

Experiences and Lessons from a Pilot Line

H. Schade and H. Maurus

Production Volume and Capacity

Over the last years the production of photovoltaic modules has shown impressive growth rates of the order of 20 %, and has reached a yearly output of around 280 MW/yr in 2000 [Maycock 2000, Photon International]. Even more impressive are published announcements for sizable expansions and new constructions of production facilities that in total would provide for a yearly production capacity of up to 500 MW/yr by 2003 [private compilation 1999]. A substantial fraction (40 %, compared to 10 % in 2000) is planned to be represented by thin-film technologies.

However, there is increasing evidence of discrepancies between capacities and actual production shipments that typically persist much longer beyond the completion date of the facility than ought to be anticipated. These situations may individually and financially prove to be more serious than announced capacities or facilities that never come to fruition. The reasons for these discrepancies are manifold and different from case to case, and often can only be suspected.

Pilot Line of Phototronics

The purpose of this contribution is to describe some of the first-hand experiences that were gathered in the course of ramping up the module production to full capacity of a pilot line. Case in point is the production facility of ASE-Phototronics for amorphous silicon (a-Si) based thin-film modules. This facility, located in Putzbrunn outside of Munich, was conceived as a pilot line with a maximal output of 1 MW/yr, and was inaugurated at the end of 1991. At that time the a-Si technology was mature enough to be considered for industrial realization, which implies large-area deposition, monolithic series connection of cells, and encapsulation. The performance of this line in terms of its yearly output in number of module plates is shown in Fig. 3.8. The main message of this graph is the relatively long time scale of the rising output. If all these plates were standard modules of 0.6 m^2 at the stabilized peak

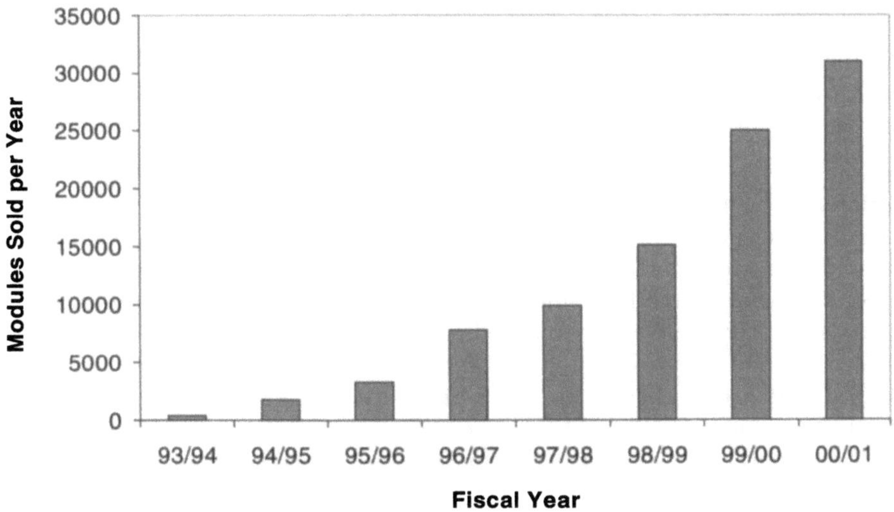

Fig. 3.8. Yearly module output

power of 32 W (i.e. 6.0 % efficiency, aperture area, as production average with commercially available TCO; see also Table 3.13), the output corresponded to close to 0.8 MW/yr by the end of 2000. Without going into detail, the substantially lower production volume than originally anticipated results from many good reasons, among these:

- lower module efficiencies than envisioned (over 10 years ago)
- lower throughput, mainly due to lower a-Si deposition rates
- a changed product mix, shifted from high-volume/low-revenue standard modules towards lower-volume/higher-revenue speciality products
- R&D tasks performed with pilot line equipment
- one-shift operation of the line until the end of 1998

There are hardly any comparisons available, except for the production facility of BP-Solarex in Virginia (USA) that was announced to have a capacity of 10 MW/yr and was reported to go on line in March of 1997 [Curry 1997]; the output as modules shipped in 2000 was 3 MW/yr [Maycock 2000].

Industrialization

Any decision to transfer research results into an industrial technology must satisfy feasibility conditions that comprise

- competitive efficiency levels
- projected low module manufacturing costs
- potential market acceptance
- financial coverage

The following considerations are derived from the specific situation at Phototronics and are based on experiences encountered there. Nevertheless, they are thought to apply more generally, and may serve as a guide and/or stimulation for improved procedures in bringing a new technology to market.

Efficiency The efficiency relevant to the product applies to the average module efficiency, referred either to the aperture area or the total module area. There are inherent reasons, as well as economic considerations, that cause the average module efficiency to be substantially lower, typically by 30 to 40 %, than the record cell efficiencies achieved at the research level. Table 3.13 summarizes the effects and the corresponding typical reduction factors that are applicable to the a-Si technology, but similarly, except for the TCO influence, will hold for other thin-film technologies as well.

The inherent losses mainly stem from the monolithic series connection of cells, i.e. both from Joule losses due to the sheet resistance of the TCO and contact resistances, and from area losses due to the cell interconnect and contact areas. Furthermore, the large-area deposition leads to areal nonuniformities, due to nonuniform potential distributions and feedgas flow distributions in the reactor for the plasma-enhanced vapor deposition (PECVD) of the a-Si layers.

Also, any production process is subject to statistics, and hence an additional inherent reduction factor arises that takes averaging into account.

Table 3.13. Typical efficiency reduction factors: from record research cell to production module.

Effect	Efficiency Reduction factor		Device	Area
	Per effect	Cumulative		
Record research cell	1.00	1.00	cell	cell
Non-optimal TCO	0.90	0.90	cell	cell
Non-uniform deposition	0.96	0.86	module	photo-active
Production-relevant processes	0.90	0.78	module	photo-active
Series resistance	0.96	0.75	module	photo-active
Cell interconnect area loss	0.96	0.72	module	aperture
Peripheral film removal	0.95	0.68	module	total
Production average	0.95	0.65	Production module	total

Specific to the a-Si superstrate technology used by Phototronics and the majority of other a-Si module producers are the constraints in the commercially available large-area TCO substrate materials that exhibit non-optimal optical properties like transparent front contact layer, compared to research-grade Asahi type U material or texture-etched sputtered ZnO [Löffel 1997]. Provided the producer has not decided to produce his own TCO, he must rely on purchased TCO material, which then must be counted as an inherent cause for reduced efficiency.

However, the choice itself for the TCO supply is a matter of judgement that is governed by economic considerations. Economic viability also determines process-related efficiency reductions: While the achievement of record cell efficiencies may rest on the use of special effects and/or processes and procedures to raise the efficiency, irrespective of the complexities and expense, the module efficiency must rely on production-relevant procedures that must satisfy throughput and yield criteria, and hence economic viability.

Table 3.13 thus shows that the cumulative effect of the described reduction factors typically amounts to over 30 %.

The module efficiency obtained in production not only determines the technical quality of the product, but it also translates into the manufacturing costs, the balance-of-system costs, further into the market acceptance, and ultimately into the financial development and outcome of the whole investment (see Sect. 4.5). Obviously, there is not a unique combination of these parameters; their interactions and trade-offs must be evaluated individually for every case.

Production-Relevant Processes The manufacturability of the solar cell structure is a necessary condition for the industrialization of a PV technology. However, to produce a finished solar module, there is additionally a large sequence of different processes that may also be unique to the particular technology and must be ready for the production environment. Figure 3.9 shows the production process flow in the pilot line at Phototronics. Ideally for the optimal utilization of a pilot or production line, the throughput for each processing step ought to be the same. However, in practice, the process requiring the most costly equipment and running with the lowest yield will be matched most closely to the throughput that is necessary to reach a certain module output in terms of square-meters per year. This particular process thus represents a "bottleneck" process, while the other process steps are left with some unavoidable reserve of allowed processing time. Here the bottleneck process is the a-Si deposition in the PECVD reactor.

Besides the throughput considerations, yield is an important issue. The resulting yield of a line is the product of the yields of each individual processing step in that line; it is directly inversely proportional to the manufacturing cost.

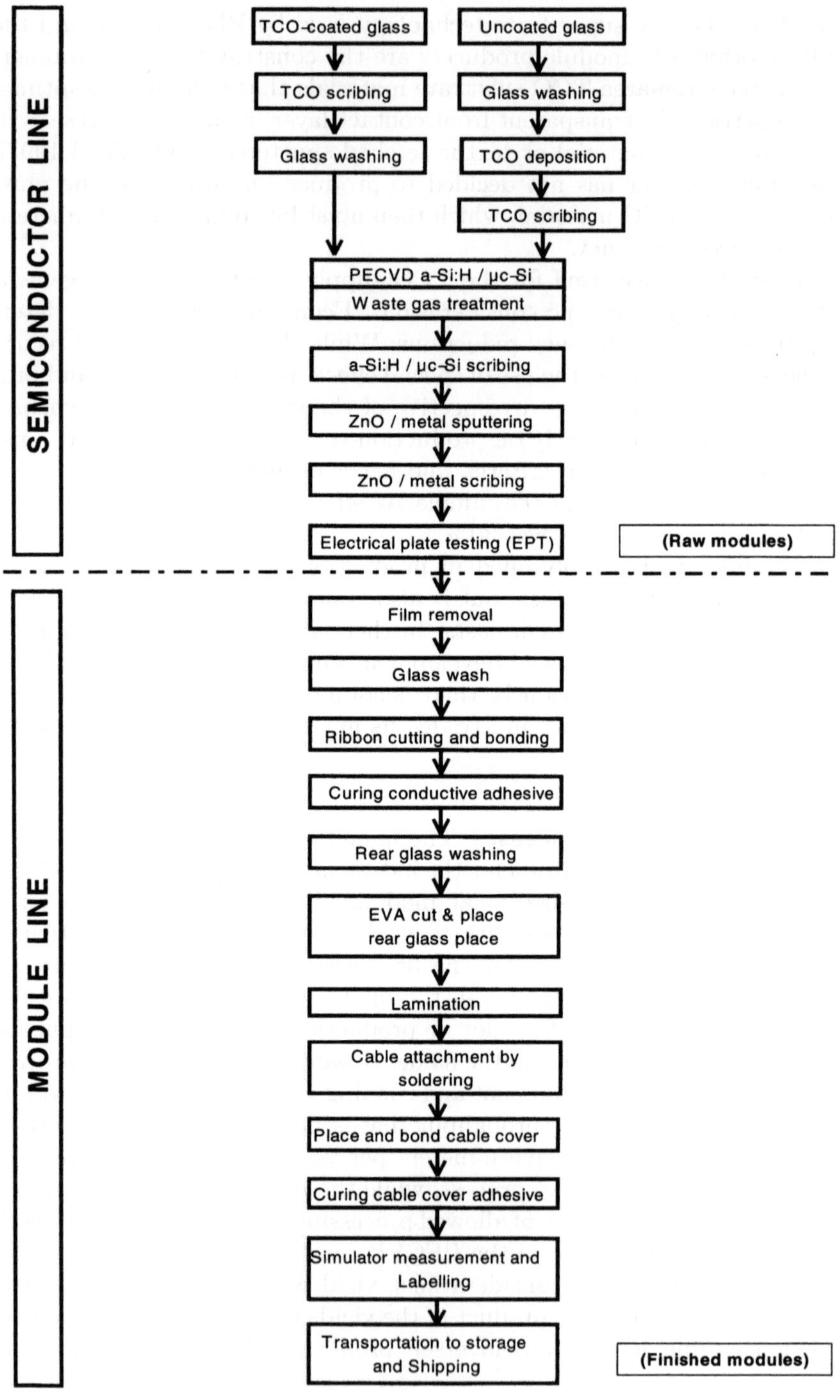

Fig. 3.9. Production process flow

As already mentioned above, many of the processing steps, apart from the a-Si deposition, are unique and cannot draw on similar applications. For example, encapsulation by producing a glass-to-glass laminate appears to be closely related to practices in the glass industry. However, an essential difference would be overlooked: the layer structure to be protected by an encapsulation is photoactive, and hence potential differences develop along the laminate. Any ingress of moisture along the laminate edges thus could lead to electro-corrosive effects. These can be avoided by film removal along the substrate edges to a certain width (i.e. sandblasting for film removal in Fig. 3.9). Also, delamination due to insufficient adhesion of the laminating foil may become an issue: the adhesion on a metal, the back electrode of the module structure, may substantially differ from that on glass. Although satisfactory adhesion is obtainable with certain metals, the metal choice might depend on the required solderability of contact leads, but still must be compatible with the adhesion requirement.

These examples serve to illustrate the complexities and interactions in other processing steps that are initially thought to be in hand or receive less attention, and then in practice could severely delay or affect satisfactory yields.

Manufacturing Costs A pilot phase is the first step in transferring successful results obtained for small-area laboratory cells to large-area devices. This technology transfer should already satisfy production relevance in terms of throughput and yield. In addition, the pilot phase should contain all features and procedures that are foreseeable in an eventual production line. The pilot phase must concentrate on the optimization of the processes and the collection of knowhow and experience. It represents an investment for the subsequent scale-up of production volume that aims at minimizing manufacturing cost by further raising throughput and yield.

An estimate of the module manufacturing costs is based on the requirement set for yearly production (in MW/yr). It includes

- investment costs, or rather equipment depreciation
- material costs
- labor costs
- infrastructure (overhead)

For a pilot line, for which many process steps still have to be firmly established, the uncertainties of the initial assumptions are obviously greater than for a further expansion of the pilot line or a full production line. The terms pilot and production line, respectively, refer mainly to the degree of experience that can be relied on, and less on the magnitude of the intended output. Accordingly, cost estimates for pilot lines may deviate more from the actual costs than would be tolerable in a generally larger production line.

A manufacturing cost estimate involves a detailed analysis of each processing step with regard to:

- required throughput (set by the required line capacity) relative to the individual processing duration; this ratio determines the corresponding equipment size or number
- materials used in the process
- yield of each process step
- equipment up-time, which is affected by both equipment maintenance and failures
- space and utility requirements for the equipment
- number of equipment operators

These data determine the direct costs. In addition, there are indirect costs that comprise administration and management, overhead, and location-dependent expenditures. Based on these inputs, the module manufacturing cost/W can be estimated or calculated once the line is running.

Products By now the module manufacturing costs at Phototronics are well established. In fact, the selling price of standard modules as a function of the cumulative sales volume follows a typical learning curve (see Fig. 3.10). However, not surprisingly, at the relatively low output of the line, on the order of 1 MW/yr, compared to crystalline Si lines on the order of typically 10 MW/yr, and in view of world market prices, the business with the pilot line cannot rely on selling only standard modules. Therefore, Phototronics decided early on to specialize in a large variety of OEM products and special requests for building integration, particularly in combination with its semitransparent module version (ASITHRU®). The key to these specialty products lies in the high flexibility of the module patterning: on the standard plate size of $(0.6 \times 1.0)\,m^2$, any module shape and size, from round watch dial sizes to

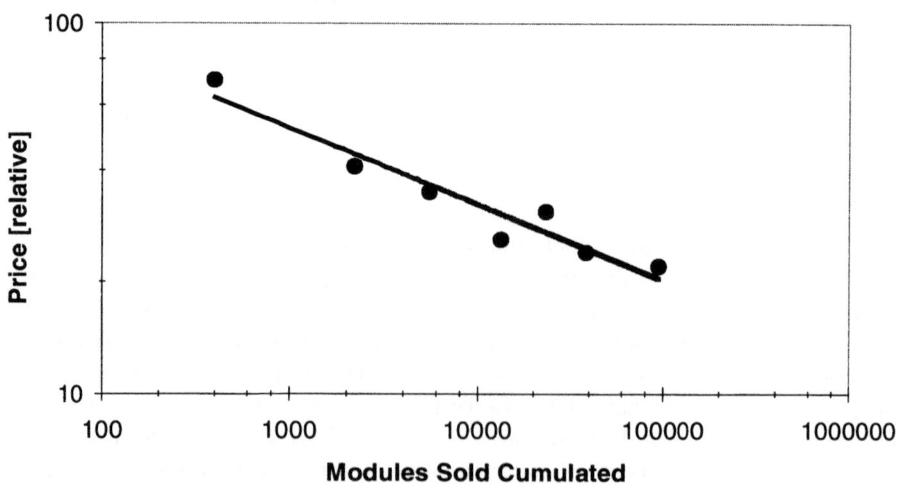

Fig. 3.10. Learning curve

custom module sizes, can be programed. Redundancy of the patterning x-y tables assures adequate throughputs at these processing stations.

Mainly by exploring these market niches, Phototronics has gradually built up its output, basically by responding to a market pull. As these markets further develop and others emerge, we will be prepared to further expand our existing line, based on today's technology, without the need for fundamental changes. Thereby, production volume of up to 3 MW/yr can be reached.

Meanwhile, developments in next-generation semiconductor deposition equipment and novel cell structures, including for example improved TCO materials [Löffel 1997] and the thin-film Si micromorph concept [Fischer 1997], are actively pursued.

Industrialization of such developments into a 10 MW/yr line may then be viewed almost as a further gradual enlargement in response to increasing market pull, and thus poses relatively small financial risk.

Financial Coverage Any business enterprise is subject to an initial investment, running costs (free cash flow), and a return on investment. Obviously, the goal is to generate a positive cash flow and thus eventually sufficient profits that allow, within a reasonable amount of time, not only to recover the investments, but also to generate sufficient capital to realize future business plans. Information on such data is generally not published.

Here, it can only be referred to an independent study of various scenarios in the build-up of a PV production line [Maisch 1999] in order to show relative changes caused by various deviations from the assumptions (see Table 3.14). The return on investment, shown here after a period of 10 years, is a measure of the profitability; positive values indicate a more or less attractive investment, negative values indicate that the investment has not yet been recovered after 10 years, and thus is not attractive.

Table 3.14. Sensitivity analysis of the return on investment affected by non-realized assumptions

Non-realized Assumptions	Manufacturing costs	Return on investment	
		in year 10	positive from year
Reference (relative)	1.00	100	7
Investment +18 %	1.06	68	8
Module eff. −10 % (rel.)	1.11	26	9
Below capacity 13 %	1.09	26	9
Module price −5 %/yr	1.00	14	9
Full capacity delayed by 2 years	1.00	10	10

The model reference is represented by an a-Si-based production line with the following characteristics:

- capacity 10 MW/yr
- total-area efficiency 7.5 %
- installation of the entire equipment, purchased "off-the-shelf," in year 0
- production ramp-up in years 1–2
- full volume at line capacity sold at market price as of year 3
- decline of module market price by 3 %/yr

The table lists the effects of non-realized assumptions, each considered individually in order of increasing impact. Accordingly, the worst case shown is a slower ramp-up with full capacity starting only in year 5. In this case, the return on investment turns positive only as of year 10, i.e. 3 years later than for the reference, and then in year 10 reaches only 10 % of the reference.

It should be mentioned that in all cases the free cash flow turns positive by year 3. However, the return on investment varies considerably, depending on the realized situation.

References

[Curry 1997] PV Insider's report, R. Curry, Ed., Feb. 28, 1997

[Fischer 1997] D. Fischer, H. Keppner, U. Kroll, P. Torres, J. Meier, R. Platz, S. Dubail, J. A. Selvan, N. Pellaton Vaucher, Y. Ziegler, R. Tscharner, Ch. Hof, N. Beck, M. Goetz, P. Pernet, M. Goerlitzer, N. Wyrsch, J. Vuille, J. Cuperus, A. Shah, Proc. 14th Europ. PVSEC, Barcelona (1997), p. 2347

[Löffel 1997] A. Löffl, S. Wieder, B. Rech, O. Kluth, C. Beneking, H. Wagner, 14th Europ. PVSC, Barcelona (1997), p. 2089

[Maisch 1999] Maisch, private communication (1999)

[Maycock 2000] P. Maycock, Ed., PV News, February 2001

[Photon International] Photon International, March 2001

[private compilation 1999] Private compilation (1999)

4 Life Cycle Assessment of Photovoltaic Systems – Trying To Fight Deep-Seated Prejudices

M. Pehnt, A. Bubenzer, and A. Räuber

4.1 Introduction

The strongest and most deep-seated prejudices against photovoltaic power generation concern ecological aspects and the energy payback time. Although these aspects are still of no relevance in the present situation due to the very low contribution of photovoltaics to total power generation, they have to be considered with reference to the importance attributed to photovoltaic power generation in long term energy economical scenarios (see Sect. 1.3.2). The latter includes photovoltaic electrical standalone off-grid systems in developing countries with a weak infrastructure, as well as grid-connected systems. In order to achieve a share of more than 1 % of the worldwide electricity supply, photovoltaics need an annual production rate of several times 10 GW_p. This means that photovoltaic power generation must be treated as a potential large-scale technology, and it must therefore be assessed ecologically like a mature power plant technology. Consequently, an integral energetic and ecological life-cycle assessment of photovoltaic power generation is already needed today.

The following aspects of an ecological assessment will be considered in this chapter:

- Primary energy consumption and energy payback time
- Global warming effect
- Other environmental and health concerns
- Availability of raw materials
- Recycling

Insofar as data are available, all PV technologies presently on the market (mono and poly silicon, a-Si, CdTe, and copper indium diselenide (CIS)) will be considered. From an energy economical point of view, not only the module, but also the complete system for electrical power generation will be assessed.

4.2 Methodology, Terms, and Definitions

4.2.1 Cumulative Energy Demand and Energy Payback Time

The question of how much energy is needed for the production of a photovoltaic power generation system and how long the system will be operated in order to recover this initial primary energy is of central importance for an assessment of photovoltaic power generation. Obviously, a power plant technology which consumes more energy in the process of being manufactured and for operation plus disposal than it produces cannot be considered sustainable. For many off-grid systems, on the other hand, the question of energy payback time is more or less irrelevant – in the same way in which energy payback times for batteries are not determined and would be meaningless.

The basic term for assessing the energy-related part of a life-cycle analysis for PV systems is the Cumulative Energy Demand (CED). The total Cumulative Energy Demand CED_{tot} contains the energy requirement – valued as primary energy – for production, installation, operation, and decommissioning of a product. In the case of most renewable energy conversion techniques, also in PV systems, the leading part of the CED is the energy requirement for production: CED_P. A clear and detailed standard specification for determining the CED for any kind of product or service is given by the German standard VDI-Richtlinie 4600 "Cumulative Energy Demand Terms, Definitions, and Methods of Calculation" [VDI 1997]. Reliable CED_P data, in particular for sophisticated products like photovoltaic systems, need a detailed analysis of process and other support energy requirements, and a list of input materials. The alternative method for determining a CED, the basically monetary Input/Output Analysis (see Sect. 4.2.2), is not sufficiently accurate for this application, and cannot discriminate, for example, between the process chains for production of different types of PV modules. However, some secondary inputs to an energy analysis such as manpower or building infrastructure can be treated on the basis of an Input/Output Analysis.

Typically, the starting point in determining the CED of a product is the final energy requirements, like electrical energy or process heat. These data must be converted into corresponding primary energy by multiplying them by the inverse conversion efficiencies of the final energy supply. For this purpose, conversion efficiencies have to be assumed. Alternatively, when data on input material are available, the corresponding heating values can be chosen to determine the primary energy demand.

Since an energy budget must obviously comply with conservation of energy, the solar energy flow into the system should in principle be taken into account. The VDI-Richtlinie 4600, however, excludes any renewable energy contribution to the CED. This procedure is obviously justified in the case of solar irradiation, since it is available without any restrictions. In the case of hydroelectric power or biomass, however, it might be appropriate to include the contribution of renewable energy in the CED because of limited availability. If CED values including a renewable energy contribution are used at all, they should only be applied in the case of a comparison between different renewable energy conversion techniques. When comparing non renewable and renewable techniques, renewable contributions to CED should be omitted.

Another term of central importance in an assessment of a PV system, or of a power plant in general, is the Energy Yield Ratio (EYR). EYR is the ratio of the annual net electrical energy output W_{el} of a power plant over its livetime LT to the total Cumulative Energy Demand CED_{tot}:

$$EYR = W_{el} \cdot \frac{LT}{CED_{tot}} .$$

In defining the EYR in this way, it should be kept in mind that this term is the quotient of two forms of energy which are not equivalent. In the definition given above, the EYR is the quotient of high-value electrical energy and typically much lower-value primary energy. From this point of view the EYR can simply be considered a kind of efficiency, such as in a power plant. Care must be taken in determining an EYR in the case of conventional power plants: as mentioned above, in this case the fossil or nuclear primary energy for operation should be included in the CED_{tot} of the power plant.

It is obvious that only renewable power plants can have an EYR greater than 1. For basic thermodynamic reasons, on the other hand, fossil or nuclear power plants will always harvest less energy than is required for operating them. Fossil or state of the art nuclear power plants therefore do not satisfy the criterion of sustainability. It is somehow puzzling that in discussions about energy recovery in renewable power plants – in particular PV power plants – this fact is hardly ever mentioned.

The most common term in this respect is the energy payback time (EPBT) of a power plant. According to the VDI-Richtlinie 4661 [VDI 2000], EPBT is defined as follows:

$$EPBT = \frac{CED_p + CED_d}{W_{prim} - CED_o} .$$

CED_p and CED_d are the cumulative energy demand for production, installation and decommissioning, respectively. CED_o is the *annual* cumulative energy demand for operation. All data exclude renewable energy contributions. W_{prim} is the annual net energy output (assessed in primary energy units) which the plant substitutes. This means that the annual net electrical output W_{el} must be converted into the corresponding primary energy W_{prim} needed to produce the same amount of electrical energy in a conventional power plant mix. This conversion is necessary because the definition of the cumulative energy demand CED analogously implies the conversion of final energies like process heat or electricity into corresponding primary energies. Omitting this conversion step in photovoltaic technology, with its high electricity demand, typically results in a severe distortion of fair and realistic assessment.

In the case of PV systems, this general definition of EPBT could be replaced simply by the quotient of the total cumulative energy demand CED_{tot} and W_{prim}, since the cumulative energy for operation CED_O is more or less negligible. However, as soon as relevant amounts of nonrenewable energy for operation must be taken into account, the general definition given above should be applied.

There remains the problem of choosing the proper power plant mix to determine the adequate conversion factor for converting electrical energy into the corresponding primary energy needed in a power plant. This conversion factor will vary according to the technical status of the power plant system – with increasing efficiency of power plants the conversion factor for Germany will decrease from more than 3 to much less than 3. Consequently, the EPBT will increase with the efficiency of the power plant system to which the PV plant is coupled. For example, a PV plant in Norway with its high share of hydroelectric power generation will have a significantly higher EPBT than in Germany (>80 % coal and nuclear power generation). As in determining EYRs, there is also the problem of determining the annual energy yield, which in solar power plants depends strongly on the location, type, and implementation of the system.

Due to these input factors, there is significant variability in determining an EPBT. Nevertheless, the EPBT of a renewable power plant is an easily understandable and unambiguously defined parameter which indicates immediately whether a technology is energy economically useful or not. However, great care must be taken in choosing the proper conditions for determining EPBT values.

We also like to emphasize that the term EPBT is not applicable to fossil or nuclear power plants. When very low EPBTs (of the order of months) are quoted for such plants, these data result from omitting the primary energy of the nonrenewable fuels for operating the plants. This procedure is unreasonable in terms of energy economics.

In the particular case of photovoltaic power plants there is a way to significantly reduce the variability in determining energy payback times: since typically 90 % or more of the process energy for producing PV modules is electrical energy, EPBT could simply be defined as the ratio of electrical process energy to annual electrical energy output. This method was applied by Nijs and co-workers [Nijs 1997] and by Knapp and Jester [Knapp 2000, Knapp 2001] who used electrically equivalent energies instead of primary energy. They thus avoided any problem of converting electrical energy into primary energy. Questions concerning the public utility power plant mix or the origin of electrical process energy turn out to be irrelevant. This procedure, however, is methodologically questionable, and its application is in any case restricted to photovoltaics. Also, great care must be taken when using the energy demand data (given in kWh electrical energy) from these authors in a comparison with truly primary energy CED data from other authors.

Finally, there remains the question of which of the indicators described above should be chosen for a meaningful energy economical assessment of PV power generation. A basic indicator given by nearly all authors dealing with the problem of energy payback time is the CED_p per m^2 or per W_p of PV modules or systems (see Table 4.2). The EPBT is a very popular indicator, but it contains no information on the remaining lifetime of the system. The EYR is an important indicator, since it assesses the complete life time of a power generation system. However, it can easily be misinterpreted, since it somehow underrates high-value electrical energy by comparing it with primary energy. In addition, it has the slight disadvantage of not being additive, which means that the EYR data of different system components cannot be added to give the EYR of the total system, as is often done in more detailed analyses of EPBT.

In trying to find the most meaningful indicator for an assessment of renewable power plants, we prefer to modify EPBT in such a way that it comprises the complete lifetime of a power plant system. We therefore define the Net Energy Life Time of a renewable energy power plant (NELT):

$$\mathrm{NELT} = \mathrm{LT} - \mathrm{EPBT}\,.$$

The NELT is directly correlated with the net energy yield of the power plant. In order to make NELT a reliable and meaningful indicator, care must be taken to include the CED of all replacement components, as well as all maintenance and repair during the lifetime of the plant.

4.2.2 The Overall Approach: Life Cycle Assessment

The calculation of the Cumulative Energy Demand (CED) is useful for a first determination of environmental weak points of systems or processes. However, it does not provide detailed information on the environmental impacts of the systems investigated because the actual environmental interventions – damage to human health, to ecosystems, and to resources – are not recorded. In many cases, the environmental impacts depend strongly on the energy used. There are many examples, however, where decisions based on the CED can lead to incorrect conclusions. For example, the production of nickel or platinum group metals from sulfidic ores leads to disproportionately high SO_2 emissions. Magnesium production involves SF_6, which has a very high global warming potential (GWP), so that the GWP is much higher than the CED suggests. To give a third example, two vehicles consuming the same amount of gasoline but with entirely different catalyst technology cause different environmental impacts, although the CED for their production and use might be almost the same.

The instrument to assess these environmental impacts is called life cycle assessment (LCA). Over the past ten years, the use of LCA has grown rapidly. Parallel to this development, an international standardization process has been started, as result of which ISO standards structure this instrument and offer guidelines for the practitioner. The two key elements of an LCA are

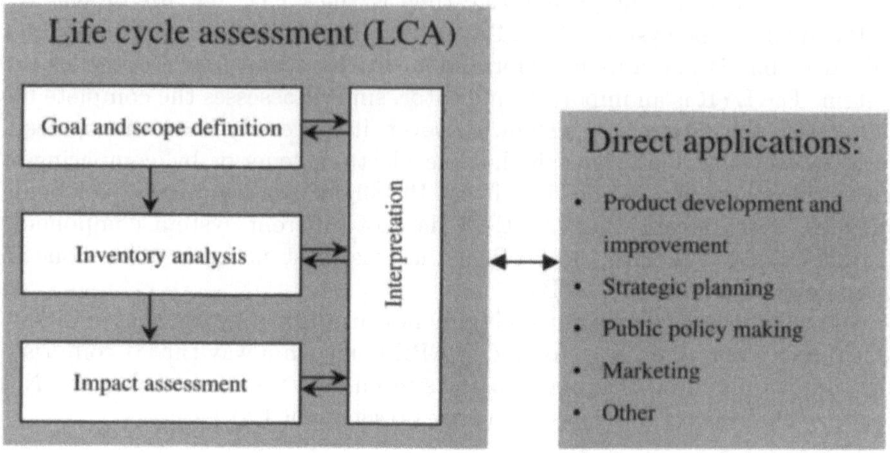

Fig. 4.1. Life-Cycle Assessment according to [ISO 14040 1997]

- assessment of the entire life cycle ("cradle-to-grave approach") of the investigated system
- assessment of a variety of environmental impacts

According to ISO, the LCA basically consists of four steps (Fig. 4.1) [ISO 14040 1997]. The first step is the Goal and Scope Definition, in which the investigated product system, the intended application of the study, the data sources, and the system boundaries are described, and the functional unit – i.e. the reference of all related inputs and outputs – is defined. The criteria for selecting input and output flows or processes must be specified. In this step, the data quality requirements, for instance time-related and geographical coverage, the consistency, typicality and uncertainty of the data, and the critical review procedure must be described. A crucial step is the determination of the investigated impact categories (see below). The Inventory Analysis (LCI) "involves data collection and calculation procedures, to quantify relevant inputs and outputs" [ISO 14040 1997]. These input and output flows involve consumed or produced goods, as well as emissions, waste streams, etc. It is essential to consider all life cycle stages, i.e. system production, operation and disposal/recycling. Typically, there will be iterative steps leading to additional data requirements. The data collection usually follows the process chain, i.e. extraction, conversion, transport, production, use, and disposal or recycling. The phases can also be divided into smaller phases, the so-called "unit processes." Each unit process of the chain has several incoming and outgoing material and energy flows, which are carefully recorded. This method of LCI is called Process Chain Analysis. While flows within the boundaries of the system are the "commodities" or "commodity flows," flows across the boundary are called the "elementary flows." The latter can be emissions (pollutants), energy carriers or other raw materials (resources). The main product or the coproducts, energy carriers, accessories, wastes, and emissions into air, water or soil are outputs which leave the process boundaries.

While the LCI via Process Chain Analysis can be considered a physical way of determining the flows within a system using rules of conservation of mass and energy within defined system boundaries, the Input/Output Analysis (IOA) solves the task with economic methods. Based on Input/Output tables, which give an aggregated description of the activities of an economy by showing monetary flows between economic sectors, an input coefficient matrix can be derived describing the demand of product inputs per unit output for each sector. The matrix elements describe the monetary flows between the sectors. Knowing the monetary value of products leaving a process, and associating them with their corresponding economic sectors, allows a calculation of the elementary flows via the sectors of demand for each supplying sector of interest.

Process Chain Analysis and Input/Output Analysis can be linked in the so-called Hybrid Approach [Marheineke 1999]. The monetary value of the commodity input not yet covered by the Process Chain Analysis can be calculated by subtracting the known inputs in monetary units and the net value added to the process from the value of the commodity output. The remainder necessarily constitutes the amount of uncovered inputs, assuming that the value of all products has been completely determined. This remainder is now assessed using average emission coefficients of the respective sectors.

The potential impacts of the inputs and outputs of Inventory Analysis are then determined in the Impact Assessment, the third step in the LCA process. The Impact Assessment categorizes and aggregates the input and output flows to the biosphere into so-called impact categories, such as the global warming, multiplying by characterization factors.

The development of impact categories with relevant characterization factors has been discussed intensively in [CML 1992], with more recent developments published frequently in the International Journal of Life Cycle Assessment. For an overview see also [Wenzel 1997]. Impact categories include

- depletion of abiotic resources, such as fossil energy carriers and uranium, metals, or other materials: In this impact category, CED (see Sect. 4.2.1) can be calculated by multiplying the amount of energy carriers required by the respective energy content. The energy content should be oriented to the amount of energy extractable with the best available technologies. Thus, the higher heating value of fossil energy carriers should be favored [Frischknecht 1998]. In most cases, however, the lower heating value is applied to determine the efficiencies.
- depletion of biotic resources as a measure of overexploitation
- global warming: The emission of greenhouse gases influences the stability of solar irradiation and absorption/reflection at the surface. These gases, such as carbon dioxide, methane, ozone, and nitrous oxide, absorb

infrared radiation emitted by the earth and thus increase the average temperature. A global warming potential (GWP) can be assigned to these anthropogenic climate gases which characterizes their effectiveness at increasing the temperature relative to carbon dioxide for a given reference time. Most recent GWPs are published by the Intergovernmental Panel on Climate Change.

- depletion of stratospheric ozone, particularly by chlorinated and brominated compounds, nitrous oxide, and indirectly by the greenhouse effect: Ozone depletion is quantified using the ozone depletion potential with CFC-11 as reference substance.
- acidification: Several substances, particularly sulfur dioxide and nitrogen oxide, act as proton sources and acidify soil and water. The impact category can be operationalized using the acidification potential, which is the ratio of the number of potential proton equivalents per unit mass of a substance to the number of potential proton equivalents per unit mass of sulfur dioxide as a reference [CML 1992].
- nutrification (eutrophication), i.e. the addition of mineral nutrients to soil and water, which results in increased algae growth, a reduction in ecological diversity, and in some instances, a lack of oxygen due to increased algae growth: Nitrogen and phosphorus components are the contributing factors to nutrification. Nutrification can be quantified as the ratio between the potential biomass per emitted substance and the potential biomass per reference substance, commonly PO_4^{3-} [CML 1992].
- emission of ecotoxic and human toxic substances, e.g. pesticides, heavy metals, and carcinogenic substances: For these complex impact categories, a number of different quantifiers have been tried [Wenzel 1997].
- emission of radioactive substances [Solberg-Johansen 1997, Frischknecht 2000]
- other impact categories, such as land use, noise, waste, and odor

According to [ISO 14042 2000], the next elements of the impact assessment are optional. They include

- normalization, i.e. the division of the environmental impacts per functional unit by reference environmental impacts (e.g. the daily impacts per capita) to gain further understanding of the magnitude of an environmental problem
- grouping, for instance sorting the impact categories on nominal or ordinal scales based on value choices
- weighting, i.e. "converting indicator results by using numerical factors" [ISO 14042 2000]. It is unavoidable that these aggregation steps are based on assumptions on the valuesphere, i.e. the perceived seriousness of ecological damages.

The fourth and last step is the interpretation, which analyzes the results, reaches conclusions, and puts forward recommendations while explaining the limitations of the study.

4.3 Life Cycle of Photovoltaic Systems

4.3.1 Important Parameters for the Cumulative Energy Demand (CED) and Life Cycle Assessment (LCA) of Photovoltaics

The environmental impacts U_j (with j environmental impact categories) of PV systems are the total of the impacts of production, use, and end-of-life impacts. The LCAs carried out in the literature, however, demonstrate that for most impact categories, such as global warming or acidification, the production of the modules is responsible for more than 90 % of the life cycle emissions. Thus, the environmental impacts can be written as

$$U_j \text{ per kWh}_{el} = \frac{U_{j,\text{Production}} \text{ per kW}_p}{\text{specific annual energy yield} \left(\frac{\text{kWh}_{el}}{\text{kW}_p a}\right) \cdot \text{lifetime}\,(a)} + \text{Rest}$$

$$= \frac{U_{j,\text{Production}} \text{ per m}^2}{\text{global solar irradiation} \left(\frac{\text{kWh}}{(\text{m}^2 a)}\right) \cdot \eta_{STC} \cdot \text{quality factor} \cdot \text{lifetime}\,(a)} + \text{Rest}$$

where η_{STC} describes the efficiency of the system under standard test conditions, and the quality factor quantifies the further reduction of system performance due to deviations from these standard conditions.

The equation and Fig. 4.2 show that the outcome of an LCA of photovoltaic energy production depends on three main determinants.

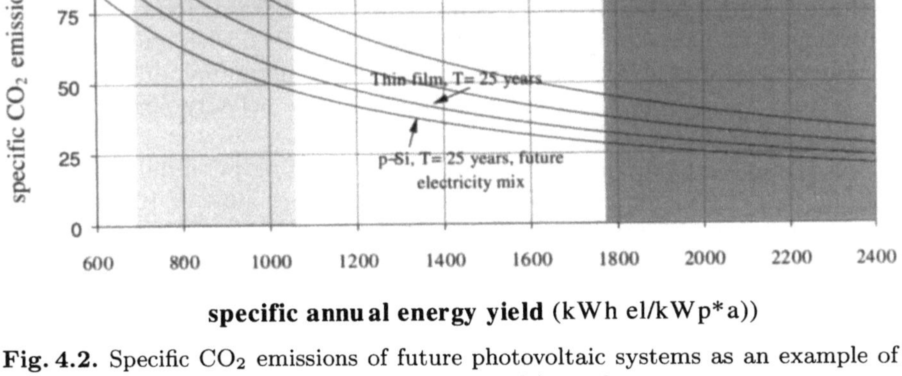

Fig. 4.2. Specific CO_2 emissions of future photovoltaic systems as an example of the influence of important parameters on the LCA results

a) *The specific energy yield.* The specific energy yield varies with
   - the geographic location of the system (in regions of high solar irradiation the specific energy yield can be twice as high);
   - possible concentration of solar irradiation or tracking systems;
   - local factors, such as system shadowing, etc.;
   - the orientation of the system: The energy yield of facade-integrated systems, for instance, which are generally mounted vertically and may be oriented toward west or east, can be reduced by almost a factor of 2; and
   - the efficiency of the PV modules and the balance of the plant
b) *The environmental impacts of the production of the system.* The production of PV modules comprises the production of the materials (e.g. solar grade Si or copper, indium or selenium precursor materials), the processing of these materials to wafers or thin films, the cell and module manufacturing, the production of the infrastructure of the system, and the installation. The impacts are particularly influenced (see below for further details) by
   - the PV material;
   - the electricity source for production of the system, especially for crystalline silicon modules; and
   - the process route and process parameters (yields, recycling, etc.).
c) *The lifetime of the system.* Generally, lifetimes of 20 to 30 years are assumed. As the impacts are directly proportional to the inverse of the lifetime, variations in the lifetime can lead to substantial changes in the result of the LCA. Only limited data are available so far on the lifetime of PV systems.

4.3.2 Example: Life Cycle of Polycrystalline (p-Si) Silicon Photovoltaic Systems

In the following paragraph, the LCA procedure is illustrated for a p-Si PV system, using new data from industrial suppliers combined with published data. The functional unit is $1\,kWh_{el}$ from a $3\,kW_{el}$ roof-integrated, grid-connected PV system. We assume the use of solar-grade Si (see below) to present a near future system.

As example impact categories, the global warming potential (GWP) according to IPCC recommendations using a reference time of 100 years and the acidification potential (AP) [CML 1992] are reported.

The production of p-Si PV systems can be subdivided into four parts:

- Production of the wafer
- Cell processing
- Module assembly
- Grid connection and support system

Process Description

Metallurgical Si (mg-Si) The base material for all types of wafers (except in the still proprietary EFG process) is metallurgical grade silicon (mg-Si). mg-Si is produced from quartz in a large-scale arc furnace at 1 200 °C by means of a carbothermal reduction process using coal, charcoal, or wood chips. In the classical Siemens process used for production of electronic grade silicon (eg-Si), mg-Si is then converted into trichlorosilane ($SiHCl_3$), TCS. TCS is purified by distillation and reacts with H_2 in an electrical furnace at 1 200 °C, giving, for example, p-Si together with $SiCl_4$. $SiCl_4$ is needed for further products. The cells evaluated in the literature have for the most part been p-Si cells based on eg-Si. For these, the electricity consumption is significantly higher than for solar grade Si (sog-Si), where the purity of the material is adjusted to the lower requirements of solar material.

Solar Grade Si (sog-Si) In the near future, the crystalline Si PV industry will no longer be able to rely on eg-Si supplied by the semiconductor industry, and will be forced to depend predominantly on some kind of low-cost solar grade silicon which is specially produced for the PV industry. Therefore, our LCA is based on one potential route to low cost solar grade Si, which we assume to be representative: the Bayer process [Block 2000]. The first step of the partly proprietary Bayer process is hydrochlorination, the formation of trichlorosilane, TCS, from mg-Si using recycled $SiCl_4$ and H_2 in a fluidized bed reactor (see Fig. 4.3). In the subsequent redistribution step, TCS is purified and then converted into monosilane (SiH_4) plus silicon tetrachloride ($SiCl_4$). The latter is fed back to the hydrochlorination process. Monosilane is pyrolytically decomposed into polycrystalline silicon plus H_2 at a temperature of 600 to 700 °C. This process works fully continuously and requires less energy than the classical Siemens process. It is expected to deliver Si feedstock for the PV industry at a selling price of 12–15 Euro/kg. The lower energy demand for this type of sog-Si is reflected in a comparatively low CED for the wafer, see Table 4.2 on page 195. In a later step, sog-Si is cast in ingots. The ingots are then cut into wafers using multi-wire saws.

An electricity consumption of 17 kWh per kilogram solar grade Si, 16 kWh for melting and an overhead for auxiliary electricity demand for air conditioning, etc., are needed for the production of the wafers [Block 2000]. The silicon losses in this process are still rather high. For a 100 cm^2 wafer (weight 8.2 g) approximately 24 g Si are required.

Cell processing The cell processing comprises a list of sophisticated processes: surface etching, the generation of a pn-junction, contacting, antireflection coating and testing. Due mainly to the high-temperature processes in diffusion and in sintering of the screen printing contact grid, cell processing causes the second largest contribution to the CED of the complete p-Si PV system. Particularly because of the high temperature

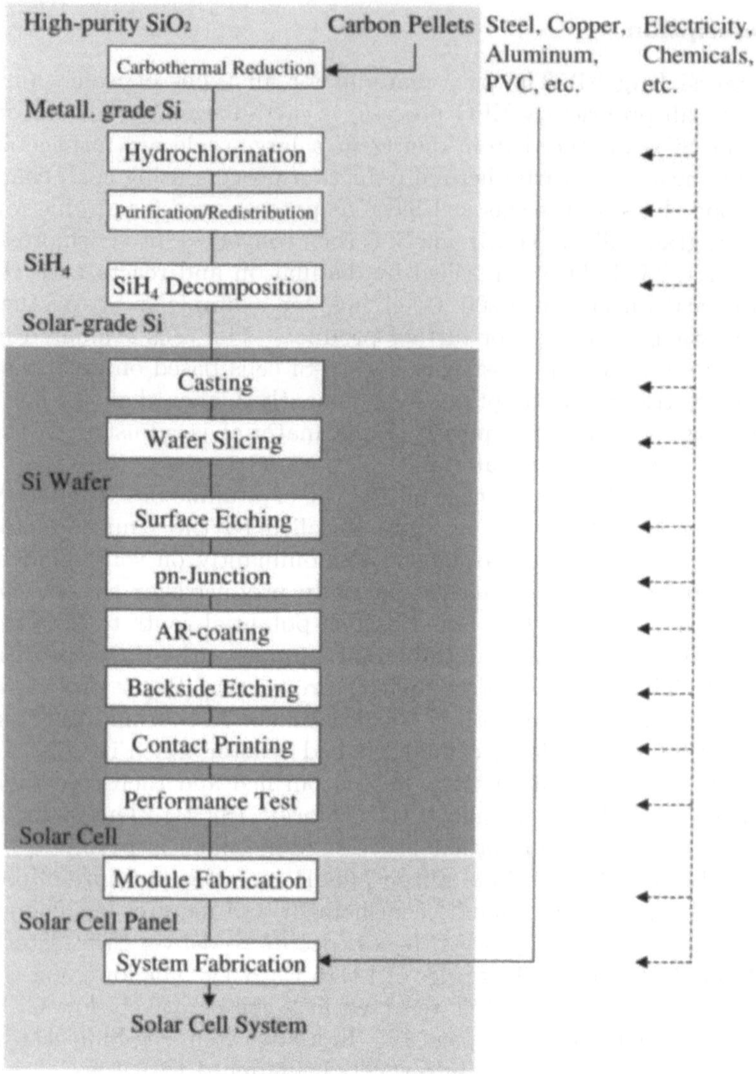

Fig. 4.3. Life cycle of p-Si PV system productions basedd on sog-Si

(800–900 °C) diffusion process, a continuous three-shift operation is essential for economical energy consumption. Consequently, the specific energy consumption for cell processing differs by a factor of 2 for continuous versus discontinuous operation [Hagedorn 1989].

Module assembly Typical process steps of module assembly are the production of cell strings, i.e. cells electrically connected in series. These strings are laminated between a front glass and a back cover glass or "Tedlar" foil using two EVA-foils. We assume a "Tedlar" foil as a back cover. The cover glass must be washed before the lamination process. Finally, an electrical contact box has to be attached to the module. A metallic frame is conventionally used in many modules. In our analysis we omit frames, since a frame is only needed as a protection during handling and mounting. In larger energy economical applications there is definitely no need for a frame. For cell and panel production, data is taken from the literature [ESU 1996, Alsema 1996].

Balance of System Among the balance-of-system (BOS) components, the array support structure has the most important impact on the LCA. The support structure accounts for more than 95 % of the specific energy requirement of the BOS components [Alsema 2000]. The DC/AC converter and electrical wiring are of minor importance. In this analysis, we assume the most favorable situation, a roof-integrated 3 kW_{el} grid-connected system on a 30 to 40 degree tilted roof in south orientation. In this case, the support structure is minimized and in addition the PV modules replace roof tiles. For the array support, aluminum/steel construction is assumed with a recycling share according to the German average.

Results The dominant impact in crystalline Si technology stems from the relatively high demand for high-purity bulk material because of the material losses of the sawing process (except for cases in which EFG technology is used). In addition, the highest process temperatures during production are around 1 400 °C due to the melting temperature of silicon. The four (high-temperature) process steps which dominate the energy demand are the production of metallic grade Si, its purification, crystallization and cell processing. No matter whether wafers are produced from single crystals, polycrystalline blocks, or large wafer plates (EFG technology), the high energy demand, typically 90 % electricity, will remain the dominant impact on LCA. Wafer production is the dominant part in terms of CED.

To characterize two different regional applications (e.g. central Europe and North Africa), locations with medium (950 kWh/kW_p) and high (1 700 kWh/kW_p) solar irradiation have been selected.

The life cycle impacts of p-Si solar cells stem mainly from three different sources (in order of decreasing importance for CO_2; Fig. 4.4):

a) impacts from direct electricity supply for the production process
b) impacts from producing the materials
c) direct emissions from the production process

Emissions from transporting and installing the systems and from their operation are almost negligible.

a) For most air and water emissions, significantly more than 20 % of the emissions are caused by the demand for conventionally produced electricity as a direct input to the PV production process. In the case of CO_2 emissions, more than 55 % are due to direct fossil electricity production.

Table 4.1. Selected environmental impacts of p-Si electricity production (using solar grade Si) in the example LCA

per kWh_{el}		Medium solar irradiation ($950\,kWh/kW_p$)	High solar irradiation ($1\,700\,kWh/kW_p$)
Non renewable resource consumption[a]			
CED (LHV)[b]	MJ	1.62	0.91
CED (HHV)[b]	MJ	1.74	0.97
Emissions			
CO_2	g	92	51
CH_4	mg	177	99
N_2O	mg	2.8	1.5
SO_2	mg	302	169
NO_x	mg	139	78
CO	mg	131	73
NH_3	mg	0.55	0.31
Particle/dust	mg	60	33
NMHC	mg	46	25
Chlorides into water	g	1.4	0.8
Chemical oxigen demand	mg	231	126
Heavy metals into water	mg	52	28
Impact Categories			
Global warming	g CO_2 equiv.	102	57
Acidification	mg SO_2 equiv.	408	228

[a] For a discussion of other abiotic resources see Sect. 4.4.
[b] The CED is determined by multiplying the input energy carriers by the lower (LHV) or higher (HHV) heating value (fossil energy carriers) and the "heat of fission" (uranium).

Therefore, to reduce environmental impacts of PV production, the reduction in electricity demand and the introduction of cleaner electricity production seems of greatest importance.
The latter can be achieved (Fig. 4.5) by installing power plants using cleaner fuels with a higher hydrogen to carbon ratio (i.e. natural gas instead of coal), by using more efficient power plant technology, by using byproducts (e.g. combined heat and power production), and/or renewable primary energy carriers.

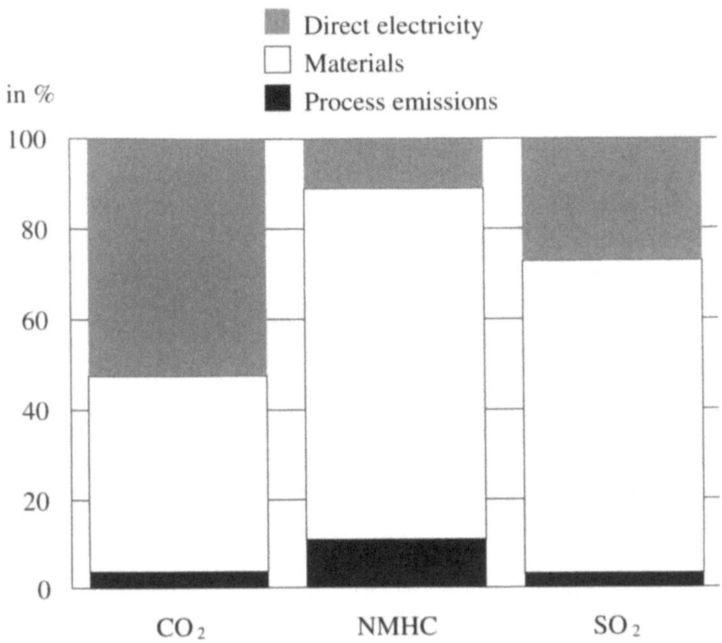

Fig. 4.4. Contribution of various emission sources to selected emissions of p-Si electricity production

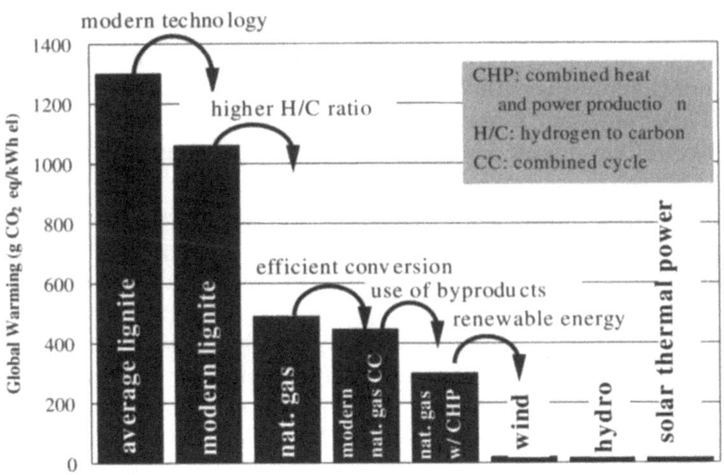

Fig. 4.5. Ways to reduce environmental impacts (example: global warming) from electricity production [Pehnt 2000]

For the next ten years, no significant changes in the CO_2 intensity of the electricity mixes of many industrialised countries are forecast due to various developments compensating one another, such as the enhanced use of natural gas, reduced nuclear energy power production, higher efficiencies, etc. In the long term, however, a substantial shift towards renewable energy systems will be necessary (see also Sect. 1.3 of this book) to meet environmental targets. Thus, a reduction in CO_2 emissions and other environmental impacts will be mandatory.

Another aspect is important in the discussion of environmental impacts due to the electricity supply for solar cell production. It seems somewhat unfair to impose burdens associated with our fossil energy system on a renewable energy conversion system. Another perspective would be to consider the lifetime electricity production of a PV system, e.g. 100 kWh/m^2 · year in areas with moderate sunshine, i.e. 2.5 MWh over the lifetime. From this electricity production, the amount necessary to manufacture the system could be subtracted, leading to a "net electricity production." The result of this perspective would be drastically reduced environmental impacts.

This consideration is not common practice in LCAs. A similar situation occurs, however, if a manufacturer produces part of the electricity consumed using renewable energy carriers. In these cases, not the electricity mix, but the supplier-specific mix should be used for assessing the product. In Freiburg, Germany, for instance, a module manufacturer uses photovoltaics and a biofuelled engine to produce the power and heat required for module assembly. In conclusion, the calculated environmental impacts based on electricity mix considerations mirror the current situation and do not take into account future developments or supplier specific conditions. They represent a maximum value for future developments.

b) The production of the materials leads to substantial environmental impacts as well. These can be, for instance, emissions from extracting the materials, from processing and purification, and from energy use for calcination, sintering, electrolytic steps, etc. In the case of p-Si, the main input material is solar-grade silicon produced from metallurgical-grade silicon.

   The production of other process inputs, such as acids, detergents, etc. is only of minor importance. The production of materials for the frame and the array support, and for the electrical components, etc. also contribute to the life cycle emissions. In the case of CO_2 emissions, for instance, aluminum is the main contributing factor to the emissions caused by materials, whereas SO_2 is mainly emitted by the production of copper from sulfidic ores and of metallurgical grade silicon. The non-methane hydrocarbon (NMHC) emissions originate from aluminum as well as plastics production.

c) In the case of p-Si, direct emissions from the PV production process – starting with sog-Si as input material – play only a minor role. The direct emissions from PV production include volatile organic compounds from solvents and screen printing the contacts, and water emissions from scrubbing the flue gas and from neutralizing etching and texturing solutions. Health and safety risks during production from HF, HNO_3, HCl and silane are only moderate and require the safety procedures common in semiconductor production technology.

It is instructive to determine the contribution of the system components, i.e. the panel, the construction frame, the AC/DC converter, and the electric components (Fig. 4.6 and Table 4.2). It can be seen that for these two selected impact categories, as for most others, silicon and wafer production are of comparable importance the production of the array support and the electrical installation. This is so because in this LCA, reduced electricity requirements for silicon and wafer production were considered. Consequently, the balance of plant gains in relative importance. From the global warming emissions caused by the wafers, one quarter stems from the supply of metallurgical grade silicon, the rest from processing to solar grade silicon and wafer production.

It must be noted that the environmental impacts from the array support structure depend strongly on the type of installation. As shown in [Frankl 1998] and [Bubenzer 1996], the area-specific primary energy content can vary between 2000 and 500 MJ/m^2 for existing and 200 MJ/m^2 for future systems, depending on the materials involved (concrete, clay, steel, etc.), the type of installation (retrofit or integrated), the orientation (flat or tilted roof, facade), and the potential functional substitution of, e.g., facade cladding or roof tiles.

Whereas in the LCAs for grid-connected systems the electrical system is of less importance, in solar home systems, which are designed specifical

Table 4.2. Primary energy demand for the production of p-Si photovoltaic systems in the LCA presented here and in the literature [Hagedorn 1989, Alsema 1996, Nijs 1997, Dones 1998, Frankl 1998, Kato 1999, Alsema 2000, Jungbluth 2000]

	Primary energy demand (MJ/m^2 module)	
	This study (tilted roof, integrated system)	Literature (1992–2000)
Si wafer	2 100	2 270–10 600
Cell processing	1 100	360–2 600 (state of the art) 200–1 100 (advanced)
Module assembly	360	350–1 150
Support structure and grid connection	1 000	2 000–500 (state of the art) 200–400 (advanced)
Total	4 560	

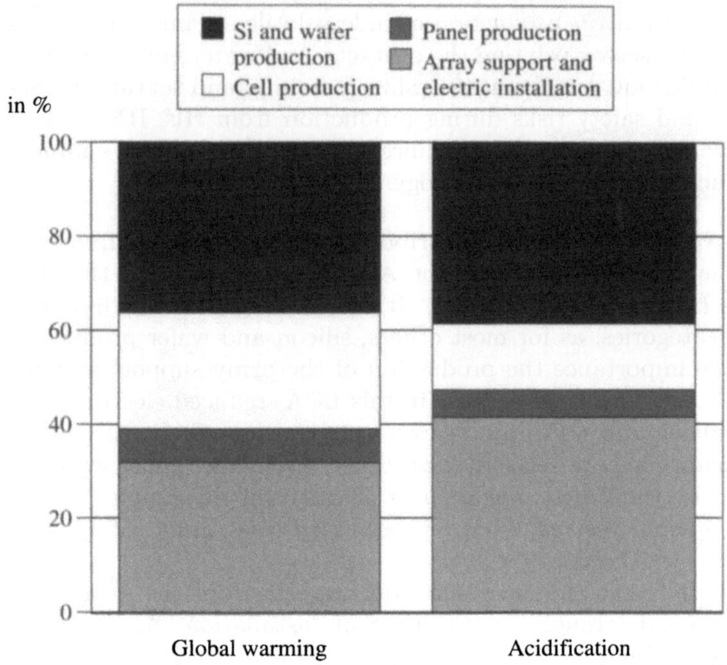

Fig. 4.6. Contribution of process steps and system components to selected environmental impacts of p-Si electricity production (example LCA with sog-Si)

for developing countries without grid infrastructure, the battery is of high importance for the overall result. In [Alsema 1998], a requirement of between 0.6 and 1.2 MJ primary energy per Wh storage capacity is calculated for the production of batteries. The relative significance of the battery is further increased by the short lifetime of that storage system. On the other hand, solar home systems should be compared to diesel generators or low voltage grid electricity of low environmental quality.

4.3.3 Aspects of LCA for Other Photovoltaic Technologies

Monocrystalline Silicon (m-Si) The life cycle for m-Si modules differs from the p-Si starting after the production of eg-Si or of sog-Si. However, the same remarks are basically valid regarding the contribution of electricity and the system components to total life cycle impacts. Wafer losses due to sawing and processing range between 50 and 60 % [Jungbluth 2000].

The production of m-Si wafers requires more primary energy than a comparable p-Si wafer, since in the first place the Czochralski process for pulling Si single crystals requires more energy than the casting process for the production of p-Si blocks. Hagedorn [Hagedorn 1989] quotes a 20 % higher energy demand for monocrystalline wafers. This trend is confirmed by the most recent and strictly empirical data on energy demand for production from the

Siemens Solar production line for m-Si modules in California by Knapp and Jester [Knapp 2000], [Knapp 2001]. The primary energy demand for Si wafer production, cell processing, and module assembly which can be derived from these data is in the upper range of the published data given in Table 4.2. In fact, monocrystalline wafers have undergone a Czochralski process twice: the eg-Si feedstock for the final Czochralski process are off-grade segments (tops and bottoms) of Si single crystals from the semiconductor industry. If this off-grade Si feedstock is regarded as a proper input to PV cell production and consequently the proportional energy consumption of the semiconductor industry process is included, extremely high values for the CED of m-Si wafers result, as in the work of Kato et al. [Kato 1998]. Eventually, however, the debate on the choice of the proper inputs to the CED for m-Si wafers will become more or less irrelevant, since in the near future the semiconductor industry source of eg-Si feedstock will only be of minor importance for PV industry. Finally, the efficiencies of m-Si modules are somewhat higher than those of p-Si modules. This reduces the required module and support structure area.

Amorphous Silicon (a-Si) The main advantage of a-Si thin-film technology in terms of LCA is its very low high-purity material requirement, about two orders of magnitude lower than in crystalline Si technology. Therefore, the environmental impacts associated with the production of electronic or solar grade Si can be eliminated. Nevertheless, in a-Si technology the dominant impact in LCAs is still the comparatively high electricity demand. The advantage of lower material requirements, however, is partially offset by the lower efficiency and thus higher balance of system requirements, and the higher degradation of the systems.

Alsema calculates CEDs for a-Si production of 17 MJ primary energy per W_p, with approximately one third from substrate and encapsulating material, one third from cell and module processing, 21 % from overhead operations and the rest from cell material and equipment manufacturing [Alsema 2000]. This is less than half the value for crystalline cells. Due to the larger required area, however – Alsema assumes 7 % efficiency for present a-Si cells compared to 13 to 14 % for crystalline silicon – the CO_2 emissions per kWh are almost identical. For future cells, Alsema assumes that the efficiency difference between thin-film and crystalline silicon decreases.

In analogy to p-Si cells, most of the air and water emissions are to an overwhelming extent due to electricity production and materials supply [Engelenburg 1994]. Thus, efforts to reduce emissions should focus on reducing the energy consumption of the processes while increasing the energy yield and lifetime of the modules. Only low emission levels are present during silane production (dust, CO_2, chlorosilanes), during glass production, deposition of the transparent conducting oxide (TCO) (tin emissions into waste water), deposition of the a-Si layers (waste gasses that are converted to solid waste in a scrubber), contacting, and module encapsulation.

The most energy-consuming processes are those in which reduced pressures/vacuum or elevated temperatures are needed over a long time. In the case of a-Si production in [Lewis 1997], for instance, the lamination process is responsible for 30 % of the total primary energy demand.

The very high contribution to the CED of a steel backplate, as assumed in the study of Hagedorn [Hagedorn 1989], is outdated. The leading technology for the production of a-Si photovoltaic modules is thin-film deposition on glass and subsequent encapsulation using a glass backplate.

Regarding environmental health and safety aspects during production, monosilane is a problematic substance. Monosilane is highly flammable and moderately toxic. Large gas clouds can build up, which can lead to sudden explosions. However, if the options for preventing hazards from silane mentioned in [Fthenakis 2000] are considered, hazards are no greater than in any other chemical industry. Dopant gases (phosphine, germane) are flammable and toxic as well, but are required in much smaller quantities.

CdS/CdTe and CIS In [ANTEC 1999], streamlined LCAs of the materials required for CdS/CdTe cells are quoted. Based on first estimates of the new production facility, ANTEC calculates a CED of 194 MJ/m^2 for the materials plus a direct electricity consumption of 144 kWh_{el}/m^2 (equaling 1 670 MJ/m^2). The resulting CO_2 emissions are 110 kg/m^2 (compared to 200 kg/m^2 for p-silicon as calculated above). The calculated energy requirements in [Kato 1999] (small-scale production: 1 800, large-scale production: 1 270 MJ/m^2 module), [Hynes 1994] (1 000 to 1 200 MJ/m^2 module) and [Alsema 1996] (450 to 1 100 MJ/m^2 module) are even lower than the values calculated by ANTEC. However, the comparison of environmental impacts per module area distorts the result because the efficiency of thin film cells is lower, and the required support structure areas greater. In [Kato 1999], for instance, the CED for CdS/CdTe module production is 41 % lower than p-Si cells per m^2 module, but the CO_2 emissions per kWh are only 29 % lower. In addition, it must be emphasized that the analyses consider target data, and should be verified once large-scale series production of CdTe cells has become common practice.

Unfortunately, [ANTEC 1999] calculates neither other emissions or impacts nor total systems. For the production of one m^2 PV substrate, ANTEC assumes approximately one kg waste, mainly defective modules which will be recycled using a procedure based on melting the modules [ANTEC 1999].

A thorough determination of CED of CIS production was carried out in [FfE 1998] for Siemens modules, leading to a CED of 3 800 to 2 400 MJ/m^2 module for pilot plant and series production, respectively. These values are substantially greater than the numbers given by [Alsema 1996] (850 to 1 700 MJ/m^2). However, the study in [FfE 1998] is more detailed, in that it uses manufacturers' data and considers all life-cycle stages. Only 15 % of the energy requirement in [FfE 1998] is due to the materials used (mainly float glass and encapsulating foil), whereas 42 % is caused by the auxiliary energy demand (e.g. air conditioning) which is particularly high because of stringent air purity requirements. In [FfE 1998], a CED for building the production equipment was also estimated, and is greater than the direct process energy demand.

CdTe and CIS cells consist of thin films containing toxic and carcinogenic Cd and other hazardous substances, such as selenium (CdTe cells), tellurium (which is less toxic than selenium), zinc, or copper. LCAs should therefore also include the potential release of these substances into the environment. For toxicity data of CdTe, CIS, and CGS see [Fthenakis 1999]. The emission of Cd, Te, and Se during the production of CIS and CdTe modules will be reduced to the minimum possible due to the health hazards associated with these materials. Particularly for Cd, emission control and continuous monitoring will be strict, whereas for Se and Te the limits are higher and the emission controls will thus be less stringent. In addition, the vapor pressure of most elements used, and thus the tendency of the material to leave the system in the gas phase, is less than or equal to 10^{-11} Torr (at 293 K). One substance which must be handled with extreme care is H_2Se, used as precursor material for CIS production.

During the mining and refining of the precursor materials, some heavy metals are released into the environment. A release of these substances could also take place when modules are crushed, when substances are leached out by rainwater, or during fires. Experimental tests for the entry of substances into water and soil have shown that even in the worst-case scenario of fractured modules and direct entry into household water or soil, the concentrations are far below (CIS) or below (CdTe) the German drinking-water regulations, and the natural abundance of an element in the soil is increased only slightly. The author concludes that only an insignificant danger to human health can be expected [Steinberger 1998b]. During a fire, the air concentrations of the relevant materials are well below critical threshold values (CIS) and below WHO and OSHA values for CdTe cells, but slightly above German threshold values for the latter. For comparison, the Cadmium intake of an infant during a fire corresponds to the effect of smoking only five cigarettes [Steinberger 1998a]. However, when large volumes of thin-film modules must be disposed of, higher concentrations of toxic materials from landfill leaching might occur. Recycling, therefore, is of increased importance not only to save input material, but also to avoid the release of toxic substances into the atmosphere.

To quantify the emissions of Cd from CdTe modules, [Alsema 1996] assumes a fire risk of 10^{-4} per annum. CdTe cells contain approximately 6 g Cd/m^2. The actual emissions released into the air depend on the type of encapsulation. It has been shown that double glass encapsulation is ad-

Table 4.3. Cd emissions due to production, operation, and decommissioning of CdTe modules [Alsema 1996]

Cd emissions (mg/m^2)	worst case	base case	best case
Mining	11	0.9	0.2
Module production	8	0.4	0.05
Utilization	1.8	1.1	0.5
Decommissioning	1.8	0.2	0.005
Total in mg/m^2	22.6	2.6	0.8
Total in g/GWh[a]	11.8	0.5	0.1
For comparison:			
Modern hard coal plant without gasification[b] (g/GWh)	16		3

[a] global average irradiation
[b] depending on filter technology and Cd content in hard coal

vantageous. Depending on certain parameters of his calculations, Alsema calculates emissions of 1.1 to 1.8 g/GWh.

Although decommissioned modules should be collected separately, a small fraction of the modules will be incinerated or disposed of at a landfill. The resulting emissions vary between 0.2 and 1.8 g/GWh.

As shown in Table 4.3, the total Cd emissions are of the same order of magnitude as those of modern coal plants, depending on worst- or best-case conditions. This shows that all measures should be taken to reduce heavy metal emissions throughout the life cycle.

In the case of CIS modules, the Se emissions from mining, production of CIS modules, utilization, and decommissioning, as calculated by [Alsema 1996], amount to 9 g/GWh in the base case (best: 2; worst: 260 g/GWh) which is one order of magnitude less than emission from a coal plant [Alsema 1996, ESU 1996].

4.3.4 Future Developments Reducing Environmental Impacts

Environmentally relevant future developments in photovoltaics will predominantly take place in module technology. In general there has been a great deal of speculation on future technological developments, so we do not go into much detail here.

Crystalline Si Modules We have integrated into our analysis one of the most important future developments, which is very close to being realized: the use of solar grade Si. However, several other developments will have similar, if not greater, impacts on LCA. The EFG method for pulling large wafer sheets from the melt eliminates all material losses, and consequently primary energy losses, from wafer sawing; a detailed LCA of this technique would be highly desirable. Reduced silicon requirements due to thinner wafers – if not compensated by lower production yields – will directly lead to nearly proportionally lower energy requirements. A higher degree of automation and more sophisticated handling equipment should make wafers of 100 to 150 μm possible, thereby reducing primary energy requirements by roughly 50 %.

Higher module efficiencies approaching 18 to 20 %, which are to be expected upon integration of high efficiency features from R&D, will likewise reduce the specific environmental impacts per kWh electrical energy from photovoltaics.

A most important (and mandatory) feature of future PV industry will be the integration of module recycling. In addition to CdTe and CIS modules, for which recycling is a must (see Sect. 4.3.3) the recycling of high-purity Si from crystalline Si modules is highly desirable, since this would significantly contribute to saving primary energy and thereby reduce CO_2 emissions and other environmental impacts. Recycling can be done at different levels [Kato 1998]: various technical solutions for recycling wafers [Wambach 1998, Frisson 1998] or even complete cells [Sakuta 1998] are available. However, a key problem of recycling will be to establish a large-scale economically feasible solution for recycling a mix of different 20 to 30 year old modules.

Thin film Modules The next step in the industrial development of any thin-film technology, which will considerably reduce specific environmental impacts, is to scale continuously running production facilities to at least 10 MW_p. As was shown by [FfE 1998] for the production of CIS modules, the main consumer of electricity during the production process is the technical infrastructure. The CED for the production equipment is also an important contribution to the specific energy requirement per kWh of photovoltaic electricity. Increasing the size and capacity of a production line has great potential for reducing these energy demands.

4.3.5 Comparison of Photovoltaics with Conventional Power Plants

To summarize the results of the preceding chapters, Fig. 4.7 gives an overview of results for regions with medium solar irradiation from LCAs in the literature for PV systems and conventional power production, as well as the data presented in Table 4.1. For regions with high solar irradiation, the appropriate values can be obtained by multiplying the emission values by the ratio of 950 kWh/kW_p to the value of the region to be investigated (e.g. 950/1 700 for North Africa). For simplicity, only greenhouse gas and acidifying emissions are shown in that figure. The global warming potential was chosen because

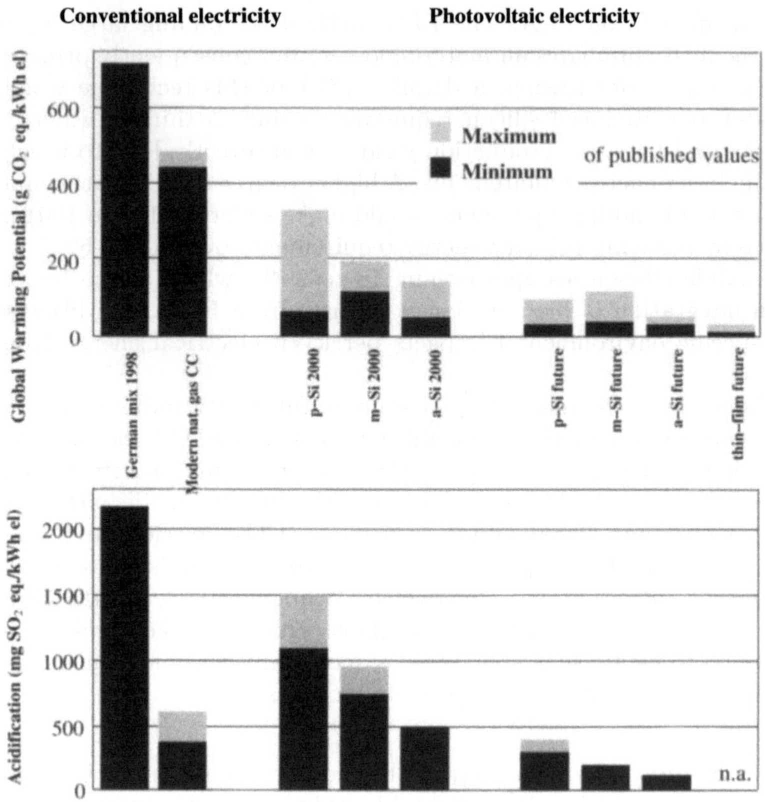

Fig. 4.7. Global warming potential (above) and acidification (below) of conventional and photovoltaic energy systems (roof-top) in regions with medium solar irradiation (950 kWh/kW$_p$). Data sources: Conventional electricity: [DLR 1999]. Photovoltaic electricity: [Dones 1996, Dones 1998, Hartmann 1998, Kato 1999, Alsema 2000, Jungbluth 2000] and this publication. "Future" means extrapolation and large-scale series production, but still conventional electricity supply for PV production

the greenhouse effect is generally considered to be one of the most crucial environmental problems, whereas acidification is shown as one example of several impact categories dominated by typical air pollutants, such as SO_2 or NO_x emissions.

Also shown in Fig. 4.7 are estimates for thin-film cells. Many of the LCAs in the literature are limited to CO_2, and energy data and do not consider other environmental impacts. In that case only the CO_2 emissions were considered for global warming potential.

It can be clearly seen that for the global warming potential, very high CO_2 reduction is possible. The high range of values for current PV systems can be explained by the high values obtained by [Hartmann 1998] who uses a hybrid method coupling the regular LCA with an Input/Output Analysis (see Sect. 4.2.2). However, Hartmann does not explain where the additional CO_2 contribution from the Input/Output Analysis originates.

Future systems, in which the advantages of large-scale series production, specially adapted materials and improvement potentials (see Sect. 4.3.4) have been realized, are characterized by even lower environmental impacts. It is important to note that for the production of these systems, the fossil-nuclear electricity mix has still been assumed. For a future renewable electricity mix these values would be even lower.

The acidification of current systems is rather high, mainly because of the electricity demand and the emissions of certain materials such as steel and copper. In addition, some of the studies assume HCl emissions from the electronic grade silicon production, which would be eliminated once solar grade silicon is used. For future systems, estimates are well below fossil power generation. In addition, efforts directed specifically at reducing the material consumption of the array support would help to reduce acidification.

Generally, LCA in the past have mainly focused on energy and CO_2 issues. Consequently, the data quality for emissions and environmental impacts other than primary energy demand and CO_2 is much poorer. In the future, detailed LCAs considering other environmental impacts should be confirmed and repeated, particularly with regard to the above-mentioned large-scale series production and concomitant realization of improvement potentials as a consequence of rapidly rising PV demand.

A crucial point in a comparison of PV systems with conventional power plants is the Energy Yield Ratio (EYR) (see Sect. 4.2.1). For thermodynamic reasons, the EYR of conventional power plants is generally less than 1. Only renewable energy power plants such as grid-connected PV systems can have EYRs much greater than 1. In terms of energy payback this means that the EPBT of a PV system must be considerably less than its lifetime.

Based on the CED data for p-Si PV systems given in Table 4.2 we have calculated the EPBT of the complete grid-connected PV systems for three different geographic zones. We assumed the current power plant mix to determine of the primary energy equivalent of the electrical energy output.

The message of Table 4.4 is simply that – wherever it is located – a grid-connected PV system with a lifetime of 25 years will harvest electrical energy for at least 20 years, free of any primary energy requirement and free of any CO_2 emissions. This means that it has a Net Energy Life Time (NELT) (see Sect. 4.2.1) of more than 20 years.

Due to the much smaller amounts of high purity materials required, and due to significantly lower process temperatures, PV systems based on thin-film technologies like a-Si or CIS have shorter energy payback times than those cited in Table 4.4 (see for example the recent study on CIS [FfE 1998]).

Table 4.4. Energy payback time (EPBT) for a polycrystalline Si grid-connected PV system

Specific annual energy yield (kWh_{el}/kW_{pa})	Energy payback time (EPBT) ($\eta_{el} = 0.315$) Int. tilted roof (years)	Net Energy Life Time (NELT) (years)
950	3.5	21.5
1 700	1.7	23.3
2 200	1.3	23.7

4.4 Availability of Materials for Photovoltaic Systems

The production of solar cell systems inevitably consumes materials, for instance the metals for the photovoltaically active layers and materials for the balance-of-plant.

4.4.1 Crystalline Si Cells

World reserves and resources of silicon are abundant (silica in various natural forms). No quantitative estimates are available in [USGS 1999], because no materials shortage can be forecast for many centuries. The only materials limitation could be silver for cell terminals (see Table 4.5 below). Depending on the geometry of the terminals, the silver demand can vary by one order of magnitude [Möller 1998]. Besides further shrinkage of the silver terminals, alternative materials for the terminals are possible.

4.4.2 Thin Film Cells

Whereas for Si solar cells no resource problems exist, concerns have been raised that the annual production rate G of thin-film photovoltaic power could be limited by the reserves of the materials. In fact, assuming that annual metal production were to be used for producing 10 % efficient solar cells, between 5 and 14 GW_p could be produced per year (the "material constrained growth" in Table 4.5), approximately 30 to 90 times the 1998 annual production. With 30 % growth in production, this value could be reached in 10 to 15 years. The following considerations are based on the thorough investigation in [Andersson 2000], with additional information from [Möller 1998] and [Zweibel 1999].

For CdTe cells, tellurium reserves are the limiting factor (Table 4.5). For CIGS, indium is scarce. The problem with metal reserves is not only the abundance of the metal in the continental crust (ruthenium, for example, is less abundant than gold, tellurium only twice as abundant), but also the scattered deposits. Gallium, for instance, is nearly as abundant as copper,

Table 4.5. Material-constrained PV expansion potentials (S: material constrained stock assuming that 1998 reserves are used to produce 10 %-efficient solar cells (100 W_p/m^2). For comparison, to meet the global annual electricity demand with PV, approximately 8 TW_p are necessary. G: material-constrained growth assuming that 1997 refinery production levels are used to produce 10 %-efficient solar cells; modified: assuming future developments (see text, from [Andersson 2000]; Ag: [Möller 1998] and [USGS 1999])

Solar cell	Metal	Metal requirement (g/m^2)	Reserves (1998) kt	Reserve/ production (yr)	S (TW_p)	G status quo (GW_p/yr)	G modified (GW_p/yr)
m-Si/p-Si	Si	No material constraint					
	Ag	2.5–33	280	17	0.8–12	50–670	
CdTe	Cd	6.3	600	30	9	310	500
	Te	6.5	20	69	0.3	5	20
CIGS	Se	4.8	70	32	1.4	46	300
	Ga	0.53	110	2037	20	10	400
	In	2.9	2.6	9	0.09	7	70
aSiGe	Ge	0.44	2	32	0.5	14	200
Dye-sensitized	Ru	0.1	6	545	6	11	20

but the copper concentrations in mineral ores are much higher. Given these rather meager available stocks, a number of approaches can significantly increase possible photovoltaic production: substitution, dematerialization, and increased availability.

Substitution involves replacing materials with functionally equivalent or similar materials, for instance replacing indium with gallium (in this case, the difference in band gap also enhances performance of the cell).

Examples of *dematerialization* include:

- reduced process losses, which can be as great as 50 % for indium or 25 % for tellurium [Zweibel 1999];
- increased cell and module efficiencies and solar irradiation (e.g. concentrator cells), thus reducing material demand per energy output;
- decreased cell thickness by means of better film quality (higher absorption coefficient) or light trapping;
- recycling defective cells during the production process.

Increased availability can be achieved by

- recycling end-of-life cells: Recycling not only reduces the energy required for PV module production by a factor of 5 [Frisson 1998], but can also enhance materials availability. For example, there are processes for recovering Si raw material, the Si wafer (with up to 75 % of the material recovered), or the entire cell. It is also feasible to use the silicon in other products, e.g. to convert silicon into ferrosilicon for steel production. For thin-film cells, recycling processes have been developed [Fthenakis 1996, Steinberger 1996];
- increased recovery: at present, the metals for thin-film photovoltaics are recovered as by-products. Increased demand and thereby increased prices – which, due to the low impact of metal prices on total production cost, are not so fatal to PV competitiveness – could promote increased recovery rates during metal mining. Indium recovery from zinc ores, for instance, could be increased dramatically. At the same time, indium could also be extracted from copper-, lead-, or tin-bearing ores. Gallium could be extracted from bauxite or germanium from coal ash;
- increased mining, including mining of deposits which are at present not economically attractive, or "exotic" deposits, such as manganese nodules on the ocean floor, which contain tellurium.

Given these stocks of resources, the available metal stocks are 2, 80 or even 150 times the reserves for ruthenium, tellurium, and indium, respectively. On the other hand, there is rising competition for metals. Indium is used, among other things, in LCD displays, alloys, and the semiconductor industry. Ruthenium might become an important catalyst material for the chemical industry or fuel cell applications. Tellurium is used in steel alloys.

Considering these factors, [Andersson 2000] recalculates the material-constrained growth rate and arrives at the revised figures in Table 4.5 (right column). It is important that metal demands for PV production be recognized early because otherwise a large fraction of the low-cost resources, especially of tellurium and indium, are currently lost in waste during mining. Another conclusion is that it is essential to develop a variety of thin film technologies, all the more so because indium, germanium, and cadmium, for instance, are all contained in Zn ores and should be exploited simultaneously to avoid the release of "unattractive," but environmentally critical metals into the environment (see Table 4.6).

Table 4.6. Conclusions from constrained materials availability (adapted from [Andersson 2000])

- Continue intensive materials science research to reduce thicknesses and increase efficiencies
- Optimize process yields
- Diversify thin-film technologies to minimize risk
- Develop efficient recycling methods and infrastructure
- Stockpile mining residues from which metals can eventually be extracted

4.4.3 Materials for the Balance of System

The consumption of other abiotic materials, e.g. for electric installation, frames, etc. is strongly dependent on the type of encapsulation, frame materials, and array support. If systems are integrated into a roof or facade, clever system integration eliminates the need for conventional building materials such as roof tiles or facade elements. These materials must be credited in a proper assessment.

Compared to conventional power production, the consumption of resources such as bauxite and iron is higher. This is due to the lower energy density of solar irradiation and the area-related energy production (i.e. the power is proportional to the area and not to the volume). A reduction in material consumption can be achieved by increased reuse/recycling of the components. In addition, if based on the total mass of consumed abiotic materials per unit electricity produced, a lignite power plant, for instance, exhibits more than an order of magnitude greater consumption due to the weight of the lignite required. It should be noted that the environmental impact associated with the supply of these materials is already included in the emission data.

4.5 Conclusions

4.5.1 Energy Payback Time

Traditionally the most widely used, although somewhat superficial, criterion for Life Cycle Assessment (LCA) of PV systems is the energy payback time (EPBT). Very high EPBTs close to the lifetime of PV systems were published in the past. This caused a great deal of confusion and spread serious doubts about whether photovoltaics are useful technology at all. A main source of confusion was very often an inadequate use of the term EPBT. EPBT is unambiguously defined as the quotient of the Cumulative Energy Demand (CED) for PV system production and disposal and its annual net energy output. Both these data must be primary energies. This means that in addition to the CED, which is a primary energy by definition, the annual net electrical energy output must also be converted into its corresponding primary energy equivalent. Since the EPBT does not contain information on overall energy balance over the total lifetime of the system, we propose the new term Net Energy Life Time (NELT). We define NELT as the lifetime of the system minus EPBT.

What makes the determination of EPBT for PV systems a difficult task is the great complexity of the CED data, particularly for the PV module production process. This is why a detailed and complete CED analysis of a modern, already operating large-scale PV module production site is not available at present. On the other hand, sufficient data are available to calculate reliable estimates for the EPBTs of PV systems. We decided that neither the largely 10 year old data from standard single crystalline nor the partly speculative data for future thin-film technologies can be the basis for representative CED and EPBT data of PV systems. We therefore chose polycrystalline wafer technology with solar grade Si feedstock to exemplify the LCA analysis. Using this technology, we arrive at an EPBT of about 3 years for a grid-connected roof-integrated PV system in central Europe, and an EPBT of 2 years for a corresponding system in southern Europe. Assuming a lifetime of 25 years, this means that the NELT for a grid-connected PV system will be between 22 to 23 years.

In any case, a modern grid-connected PV system will supply electricity for more than 20 years without any hidden energy demand and without any hidden CO_2 emissions.

4.5.2 Life Cycle Assessment (LCA)

EPBT can only serve as a preliminary indicator of weak points in an LCA for PV systems. An LCA in fact comprises many more aspects, such as danger to human health, ecosystem quality, and resource depletion. Global warming, as one example of an important environmental impact, is mainly due to CO_2

emissions, which largely stem from the relatively high electricity demand for production of any type of PV module. As in our analysis of EPBT, we have chosen p-Si PV technology based on solar grade Si as a representative system. It can be clearly shown that a grid-connected PV system considerably reduces CO_2 emissions below those of a conventional power plant mix. However, it should be kept in mind that these emissions are due to the baggage of fossil energy consumption imported into the renewable energy system.

The specific CO_2 emission per kWh photovoltaic electrical energy for a representative p-Si PV system in middle Europe is less than 20 % of the specific emissions of a typical conventional power plant mix.

Significant dangers to human health and safety due to production, operation, and disposal of PV systems do not exist in any of the Si-based PV systems, including amorphous Si. Potential impacts in this field could show up due to Cd in CdTe modules or Cd and Se in CIS modules. If, however, we compare these potential emissions with the Cd and Se emissions from a coal-fired power plant, it turns out that, assuming a certain fire hazard, Cd emissions are of the same order of magnitude and Se emissions are less than 10 % of the emissions of a coal-fired power plant. Nevertheless, controlled recycling or decommissioning of CdTe PV modules in particular and a reduction in upstream heavy metal emissions should be mandatory.

Other emissions which are potentially harmful to human health and the environment originate from electricity supply as well as materials production. Future developments such as reduced energy demand for PV production, cleaner electricity supply mixes, recycling, large-scale production allowing even stricter emission controls, and other considerations will contribute to further reduction of such environmental impacts. It might even be that materials such as indium and tellurium, which at this time are only a byproduct of mining and partly disposed of as waste, can turn out to be economically valuable materials for PV production.

Standard Si-based PV systems do not cause significant dangers to human health and safety. Other impact categories, such as acidification and eutrophication, will be significantly lower for future PV systems than for conventional power plants. Potentially dangerous emissions such as Cd from CdTe and Se from CIS modules do not exceed corresponding specific emissions from standard coal-fired power plants. Future developments will lead to a further decrease in life-cycle emission.

Potential resource depletion due to greatly increased use of mineral resources for the production of PV systems does not impose any constraints on the expansion potential of Si-based PV technology. Material constraints might turn up for Te used in CdTe, and particularly for In used in CIS. Since, however, 100 % recycling of these materials might in any event be mandatory for safety reasons, this problem may not arise at all.

Material-constrained expansion of PV technology can only occur in CdTe and CIS technology, if it occurs at all. However, the limiting production volume for potential material shortage is about two orders of magnitude above present total worldwide volume.

4.5.3 General Issues

The LCA of photovoltaic systems is still incomplete. New and complete data from modern large-scale production facilities is needed. In modern large-scale facilities, environmental issues must be taken rigorously into account. Specifically, this means that electricity consumption must be drastically reduced, and that heat recovery and recycling of byproducts during the production process should be implemented. Finally, recycling of all types of modules must become common practice. For crystalline Si modules, this will significantly reduce the CED for production, and for CdTe and CIS modules it is mandatory for safety and for conservation of rare materials.

References

[Alsema 1996] E. A. Alsema: Environmental Aspects of Solar Cell Modules. Report no. 96074, ISBN 90-73958-17-2. Utrecht University, Dept. of Science, Technology and Society, Utrecht 1996.

[Alsema 2000] E. A. Alsema: Energy Pay-back Time and CO_2 Emissions of PV systems. Prog. Photovolt. Res. Appl. 8 (2000), p. 17–25.

[Andersson 2000] B. A. Andersson: Materials Availability for Large-Scale Thin-film Photovoltaics. Prog. Photovolt. Res. Appl. 8 (2000), p. 51–76.

[ANTEC 1999] ANTEC: Ganzheitliche ökologische Bewertung der CTS-Technik nach dem ANTEC-Verfahren. debis Systemhaus Industry GmbH, T. Prinz, Stuttgart 1999.

[Block 2000] Personal Communication Bayer Solar, Dr. Block, Leverkusen 2000.

[Bubenzer 1996] A. Bubenzer, J. Butz, M. Edel, U. Vidal and A. Zimmermann: Ein universelles Dachintegrationssystem für rahmenlose PV-Module. Elftes Symposium Photovoltaische Solarenergie, Staffelstein (1996), p. 523–526.

[CML 1992] CML, R. Heijungs, J. B. Guinee, G. Huppes, R. M. Lankreijer, H. A. U. d. Haes and A. W. Sleeswijk: Environmental Life Cycle Assessment of Products. Guide and Backgrounds. Center of Environmental Science, Leiden 1992.

[DLR 1999] H. Dienhart, M. Pehnt and J. Nitsch: Analyse von Einsatzmöglichkeiten und Rahmenbedingungen verschiedener Brennstoffzellensysteme in Industrie und zentraler öffentlicher Stromversorgung. Untersuchung für das Büro für Technikfolgenabschätzung beim Deutschen Bundestag. Deutsches Zentrum für Luft- und Raumfahrt (DLR) e.V., Institut für Technische Thermodynamik, Stuttgart 1999.

[Dones 1998] R. Dones and R. Frischknecht: Life-cycle Assessment of Photovoltaic Systems: Results of Swiss Studies on Energy Chains. Prog. Photovolt. Res. Appl. 6 (1998), p. 117–125.

[Dones 1996] R. Dones, U. Gantner, S. Hirschberg, G. Doka and I. Knoepfel: Environmental Inventory for Future Electricity Supply Systems for Switzerland. Paul-Scherrer-Institut, ETH Zürich, Villigen, CH 1996.

[Engelenburg 1994] B. C. W. v. Engelenburg and E. A. Alsema: Environmental Risks of Amorphous Silicon Solar Cell Modules. 12th European Photovoltaic Solar Energy Conference, Amsterdam (1994), p. 305–308.

[ESU 1996] ESU, Ökoinventare von Energiesystemen. Grundlagen für den ökologischen Vergleich von Energiesystemen und den Einbezug von Energiesystemen in Ökobilanzen für die Schweiz. R. Frischknecht et. al. Zürich 1996.

[FfE 1998] A. Ampenberger, E. Hellriegel and D. Köhler: Massen- und Energiebilanzen für die Herstellung von CIS-Dünnschichtsolarmodulen. Forschungsstelle für Energiewirtschaft, München 1998.

[Frankl 1998] P. Frankl, A. Masini, M. Gamberale and D. Toccacelli: Simplified Life-cycle Analysis of PV systems in Buildings: Present Situation and Future Trends. Prog. Photovolt. Res. Appl. 6 (1998), p. 137–146.

[Frischknecht 2000] R. Frischknecht, A. Braunschweig, P. Suter and P. Hofstetter: Human health damages due to ionising radiation in Life Cycle Impact Assessment. Environmental Impact Assessment Review 20 (2000).

[Frischknecht 1998] R. Frischknecht, R. Heijungs and P. Hofstetter: Einstein's Lessons for Energy Accounting in LCA. Int. J. LCA 3 (1998), p. 266–272.

[Frisson 1998] L. Frisson, H. Hofkens, K. d. Clerqk, J. Nijs and A. Geeroms: Cost Effective Recycling of PV modules and the Impact on the Environment, Life Cycle, Energy Payback Time and Cost. Proceedings of the 2nd World Conference on Photovoltaic Energy Conversion, Vienna (1998), p. 2210–2213.

[Fthenakis 1996] V. M. Fthenakis, C. Eberspacher and P. D. Moskowitz: Recycling Strategies to Enhance the Commercial Viability of CIS Photovoltaics. Progress in Photovoltaics 4 (1996), p. 447–456.

[Fthenakis 1999] V. M. Fthenakis, S. C. Morris, P. D. Moskowitz and D. L. Morgan: Toxicity of Cadmium Telluride, Copper Indium Diselenide, and Copper Gallium Diselenide. Prog. Photovolt. Res. Appl. 7 (1999), p. 489–497.

[Fthenakis 2000] V. M. Fthenakis and P. D. Moskowitz: Photovoltaics: Environmental, Health and Safety Issues and Perspectives. Prog. Photovolt. Res. Appl. 8 (2000), p. 27–38.

[Hagedorn 1989] G. Hagedorn, H. Kuhn, S. Lichtenberger: Kumulierter Energieverbrauch für die Herstellung von Solarzellen und photovoltaischen Kraftwerken, Forschungstelle für Energiewirtschaft, München (1989).

[Hartmann 1998] D. Hartmann and M. Kaltschmitt: Von der Wiege bis zur Bahre. Vergleich der gesamten Emissionen einer Stromerzeugung aus regenerativen Energien. BWK 50 (1998), p. 56–61.

[Hynes 1994] K. M. Hynes, A. E. Baumann and R. Hill: An Assessment of the Environmental Impacts of Thin Film Cadmium Telluride Modules Based on Life Cycle Analysis. Proc. First World Conference on Photovoltaic Energy Conversion, Hawaii (1994), p. 958–961.

[ISO 14042 2000] ISO: Environmental Management – Life Cycle Assessment – Life Cycle Impact Assessment. 2000.

[ISO 14040 1997] ISO-14040: Environmental Management – Life Cycle Assessment – Principles and Framework 1997.

[Jungbluth 2000] N. Jungbluth and R. Frischknecht: Literaturstudie Ökobilanz: Photovoltaikstrom und Update der Ökobilanz für das Jahr 2000. Report for the Bundesamt für Energie, Projekt-Nr. 39489, Uster (Switzerland) 2000.

[Kato 1998] K. Kato, A. Murata, K. Sakuta and K.Yamada: A possibility of recycling PV-modules from a life cycle point of view. 2nd World Conference and Exhibition on Photovoltaic Solar Energy Conversion, Vienna (1998), p. 1953–1958.

[Kato 1999] K. Kato: Energy Resource Saving and Reduction in CO_2 Emissions as Values of PV Technology – A review on life cycle analysis on PV technologies in Japan. IEA/PVPS Task 1 workshop "Added value of PV system", Sapporo (Japan).

[Knapp 2000] K. E. Knapp and T. L. Jester: An Empirical Perspective on the Energy Payback Time for Photovoltaic Modules. 28th IEEE Photovoltaics Specialists Conference, Anchorage, Alaska, 2000.

[Knapp 2001] K. E. Knapp and T. L. Jester: Empirical Investigations of the Energy Payback Time for Photovoltaic Modules. Solar Energy, to be published, 2001.

[Lewis 1997] G. Lewis and G. A. Keoleian: Life Cycle Design of Amorphous Silicon Photovoltaic Modules. National Risk Management Research Laboratory, Cincinnati (USA) 1997.

[Marheineke 1999] T. Marheineke, W. Krewitt and R. Friedrich: Ganzheitliche Bilanzierung von Stromversorgungstechniken nach heutigem und zukünftigem Stand der Technik mit geografischem Bezug Deutschland. Entwicklungslinien in der Kraftwerkstechnik, Essen, VDI-Bericht, 1999.

[Möller 1998] J. Möller: Integrierte Betrachtung der Umweltauswirkungen von Photovoltaik-Technologien. Diploma thesis, University of Oldenburg, Oldenburg 1998.

[Nijs 1997] J. Nijs, R. Mertens, R. v. Overstraeten, J. Szlufcik, D. Hukin and L. Frisson: Energy Payback Time of Crystalline Silicon Solar Modules. Advances in Solar Energy 11 (1997), p. 291.

[Pehnt 2000] M. Pehnt: Life Cycle Assessment of Renewable Energy Technologies. In: Key Technologies: Renewable Energy Technologies. Report for the HGF Project: Sustainability. Perspectives for Germany. DLR, Institut für Technische Thermodynamik, Stuttgart 2000.

[Sakuta 1998] K. Sakuta, K. Otani, A. Murata, H. Unagida and K.Kurakawa: Attempt to recover Silicon PV-cells from modules for recycling. 2nd World Conference and Exhibition on Photovoltaic Energy Conversion, Vienna (1998), p. 2268–2271.

[Solberg-Johansen 1997] B. Solberg-Johansen, R. Clift and A. Jeapes: Irradiating the Environment: Radiological Impacts in Life Cycle Assessment. Int. J. Life Cycle Assessment 2 (1997), p. 16–19.

[Steinberger 1996] H. Steinberger, A. Kettrup, W. Thumm and P. D. Moskowitz: Umwelt- und Gesundheitsauswirkungen der Herstellung und Anwendung sowie Entsorgung von Dünnschichtsolarzellen und Modulen. Report to the Statusreport 1996 Photovoltaik. Forschungszentrum Jülich, Jülich (Germany) 1996.

[Steinberger 1998a] H. Steinberger: Health and Environmental Risks from the Operation of CdTe- and CIS-Thin Film Modules. ND World Conference and 2nd Exhibition on Photovoltaic Solar Energy Conversion, Vienna (1998), p. 2276–2278.

[Steinberger 1998b] H. Steinberger: Health, Safety and Environmental Risks from the Operation of CdTe and CIS Thin-film Modules. Prog. Photovolt. Res. Appl. 6 (1998), p. 99–103.

[USGS 1999] USGS: Mineral Commodity Summaries. U. S. Geological Survey, 1999.

[VDI 1997] VDI-Richtlinie 4600: "Cumulative Energy Demand, Terms, Definitions and Methods of Calculation", VDI-Handbuch Energietechnik, 1997.

[VDI 2000] VDI-Richtlinie 4661: "Specific Numbers for Energy Assessment- Definitions – Terms – Methodology", VDI-Handbuch Energietechnik, draft, 2000.

[Wambach 1998] K. Wambach: Recycling of PV-modules. 2nd World Conference and Exhibition on Photovoltaik Energy Conversion, Vienna (1998), p. 2248–2251.

[Wenzel 1997] H. Wenzel, M. Hauschild and L. Alting, Environmental Assessment of Products. Volume 1 and 2. Chapman Hall, London 1997.

[Zweibel 1999] K. Zweibel: Issues in Thin Film PV Manufacturing Cost Reduction. Solar Energy Materials and Solar Cells 59 (1999), p. 1–18.

5 Strategies in Photovoltaic Research and Development – R&D and History

A. Räuber

Most of the material presented here was collected under Research Grant 0329727 from the German Ministry of Economics and Technology (BMWi). The author soley is responsible for the content of this paper.

5.1 Introduction

5.1.1 Structure of this Chapter

The intention of this chapter is to present an historical outline of the efforts of governments, academia, and industry in research and development in the field of photovoltaics. It describes the background and motivation for activities, the funding of the efforts, and the strategies and organization of R&D in countries, rather than the technical issues.

It is not possible here to present data on the development in all countries that are engaged in photovoltaics. Only a small selection of examples is described to some extent, but it is believed that this selection is representative the situation and shows the differences in approach.

5.1.2 Photovoltaics – a Challenge for Governments, Society, and Industry

National Policies Justification for spending public money for PV R&D is very similar in various countries, but the driving forces for political action are certainly different and are subject to change under different political situations.

Demands of Society In most developed countries society has a very positive attitude towards renewable energies and especially photovoltaics. People are aware of the problems of the dependence on only a few, and in most cases imported, sources of energy, they feel at least uncertain about the safety of the use of nuclear power and of the possibility of climatic changes, and they care for a clean environment. They require action from their governments.

Opportunities for Industry In the beginning there was an remarkably broad certainty that photovoltaics would develop very rapidly into a huge market, and many big companies invested in the development of this new technology. After many years of disappointment, a number of companies went out of the field. Today, a real worldwide market has in fact started to grow, and the hope of a commercial opportunity is much better than 20 years ago. The challenge for industry is the big cut in fabrication cost that is still needed for the whole chain of components for PV systems, and the development of innovative low-cost manufacturing technologies is a must. For industry this is an investment in the future, and companies need a lot of patience to finally succeed.

Value of Solar Electricity The motivation for the development of photovoltaic electricity generation changed over the years, since a number of new aspects came up step by step:

- The first period was characterized by the extremely high price of solar cells. Only very special applications could be envisioned, and solar arrays found their way into electricity generators for spacecraft and for measuring devices at very remote places. Nobody connected PV with the idea of bulk electricity production in industrialized countries; only small standalone systems for remote dwellings in undeveloped areas were taken into account.
- The second phase set in with the energy crises of the 1970s: Development of renewable energies might be a means to improve the security of the national energy supply. Somewhat later the added value of being "clean" energy, saving the environment and the climate, was a new aspect for the promotion of photovoltaics.
- In the third phase it was realized that the development of a new energy technology means the development of a new innovative industry, which is basically a regional industry, creating jobs and including the chance for exports.

Today, the motivation for investments in the development of photovoltaics is a mixture of all the aspects mentioned. The industrialized countries see very good prospects for a new industry, and all have also in mind a huge but undeveloped market in remote areas of countries in the sun belt.

5.2 Chronology of Development

5.2.1 First Silicon and Thin-Film Solar Cells

Due to its leadership in solid state physics and especially in semiconductor physics and technology in the 1960s and 1970s, the United States is also the birthplace of solar cells of nearly all kinds. Main events in the initial phase

Table 5.1. Early events in the development of solar cells

Year	Event	Efficiency	Reference
1954	First useful silicon solar cell	6 %	Chapin 1954
1954	First cadmium sulfide solar cell	6 %	Raynolds 1954
1956	First gallium arsenide solar cell	6 %	Jenny 1958
1958	First satellite (Vanguard I) with PV solar array (108 cells, worked for 8 years)		
1959	Satellite Explorer VI with 15 W array (8 000 cells)		
1972	First cadmium telluride solar cell	5 %	Bonnet 1972
1974	First monocrystalline	12 %	Wagner 1974
1975	First CIS thin-film solar cell	6.6 %	Kazmerski 1975
1976	First (a-Si) p-i-n solar cell	2.4 %	Carlson 1976
1977	Description of light-induced degradation		Staebler 1977

are summarized in Table 5.1. It is easy to recognize that most materials for currently interesting solar cells were already there in the early 1970s.

During the first 20 years, silicon solar cells found their application mainly in space arrays on satellites, by 1975 NASA used about 1 million solar cells per year.

Up to the late 1970s most activities in thin-film development were devoted to the copper sulfide/cadmium sulfide cell, and were abandoned due to severe stability problems when other thin-film materials came up.

5.2.2 Current Status

Today's situation has been extensively described in the technology chapter. It is characterized by a fully industrialized crystalline silicon solar cell technology, which is the workhorse of photovoltaics, and several thin-film technologies that are in their starting phase. Amorphous silicon (a-Si) cells have been in pilot production for several years, CIS cells have been in pilot production since 1998 at one company, and the pilot production of cadmium telluride cells started up at the end of 2000.

5.3 Role of the State in Research and Development

5.3.1 Legislation

A wide variety of reasons for the involvement of governments is found in the various countries that have PV R&D programs. Some have just modest research projects funded through scientific research organizations, others have well-developed long-term R&D programs with the engagement of the scientific community and industry, and the third group even has a legislative basis for the enforced introduction of renewable energies in their countries. In the latter case, the governments normally take care of the program by installing appropriate administrative entities and securing long-term budgeting.

Legislation in the U.S.A. During the Carter administration in the 1970s, the oil crisis and growing awareness of environmental problems laid the basis for a new energy policy in the United States, and for a strong push towards the development of renewable energy. This movement resulted in the first Federal law for accelerated development of photovoltaics, the "Solar Photovoltaic Research, Demonstration, and Development Act" of 1978. This law authorized the expenditure of $1.5 billion by the Government within 10 years, and it was the basis for the famous FSA Project (see below), which lasted until 1985 in spite of severe budget cuts by the Reagan administration in the early 1980s.

A second important law of the year 1978 was the "Public Utility Regulatory Policies Act," called PURPA, which regulated the relationship between utilities and their customers with "cogeneration", i.e., of customers producing part of their electricity themselves. The law ensured that the utilities had to allow the connection of other generators to the grid, they had to supply backup power, and they had to buy excess power produced by the cogenerators at reasonable rates ("avoided cost basis"). PURPA was the first law in this direction, and provided the basis for grid-connected PV systems. PURPA was duplicated in various ways in many countries.

The time of the Reagan administration was characterized by a policy of noninterference with market forces. This excluded the funding of commercialization of renewable energies, and reduced the involvement of the government in the funding of basic research.

In the early 1990s President Bush slowly changed this policy, and a new national energy policy was developed ("Energy Policy Act" of 1992). From this time on, government support of the private sector to accelerate PV commercialization was again possible. In fact, this policy continues up to present, moderately enforced by "initiatives" due to President Clinton, which however do not have the force of laws. The best known is the "Million Roofs Initiative," which calls for the installation of both thermal collector systems PV systems on private homes and on public buildings. One very positive outcome of this initiative is the generation of procurement programs by a number of federal agencies for PV systems on government buildings, in national parks, and in other public places.

For the market development in the U.S., the legislation of individual States is very important. Many states offer tax credits for the installation of PV systems in businesses and private properties, and the regulations on buy-back rates for electricity generated in grid-connected systems is very important for their economic value. So-called "net metering" with equal rates for buying and selling electricity is a very common model in many states.

Legislation in Japan The Japanese Government has a long-term commitment to develop renewable energies, notably photovoltaics, and it certainly has the most consistent program. The basis of the program is a law enacted in 1974 by the Diet, the Japanese parliament.

The Sunshine Project was established for the development of energy technologies of all kinds, including renewable energies. In 1993 the much broader New Sunshine Project followed, which also included energy conservation technologies and environment issues. All these efforts in Japan are backed by legislative measures, and the administration is obliged to act according to rules given by the programs.

Situation in the European Union In most European countries, including Germany, France, and Italy, R&D in the field of renewable energies is based on initiatives from the administration rather than from legislative forces. The same is true for the European Commission. In most cases, Research Plans are worked out and stand for some general rules for the direction of R&D, but the success of the programs depends very much on the people in the respective administrations. Under these circumstances it is difficult to develop long-term strategies.

The situation is quite different for the market introduction programs of some countries in the European Union. As has been pointed out, legislation is needed to regulate the connection of dispersed electricity generators to the public grid and to ensure fair buy-back rates. The German Electricity Feed-in Law ("Stromeinspeisungsgesetz") and the Renewable Energy Law (EEG) of the year 2000 are good examples for the direction of an effective legislation. Several other countries like Spain and Italy have similar laws.

Although the Renewable Energy White Paper by the European Union requires some action at the European level, the European Commission decided to leave measures for market introduction under the legislation of the national governments for the next few years.

5.3.2 Financial Engagement

General Remarks Special budgets for the development of PV technologies appeared in a number of countries in the early 1970s. They grew rapidly, and their development over the years quite nicely reflects the changing attitude towards energy policy in the various countries. The three biggest players in the field, the U.S.A., Japan, and Germany, have each invested between \$1 and 2 billion over the years, and it is very interesting to analyze how this money has been spent.

At first glance, the amount of money spent for R&D in photovoltaics seems to be appropriate, but in reality it is very, small in view of the claim of this technology to be one of the most important options to solve our energy problems. Other long-range options like fusion receive an order of magnitude greater supportfrom most governments.

There is general agreement that PV has technically developed to a stage that enables introduction to the market, and strong programs in this direction have been introduced, by the German and the Japanese governments, but also by other countries like Spain and the Netherlands. It is, however, dangerous to introduce such programs at the cost of R&D efforts. PV is far from a mature technology. There is still room for considerable improvement of the product and of great cost reduction by new innovative technologies, which can only be developed in a joint effort comprising science and industry.

Comparison of R&D Budgets All budget figures given in this chapter have been converted to US dollars for comparison, and current US dollars have been used without taking account of inflation. This is somewhat arbitrary, since the exchange rate between a national currency and US dollars does not always reflect the real purchasing power of a currency.

U.S.A.: The budget of the U.S. Department of Energy (DOE) is shown in Fig. 5.1a. It is characterized by strong growth in the 1970s under President Carter, with a peak at $152 million in 1980, a decline in the Reagan era down to a minimum of $35 million, and a recovery initiated by President Bush in the early 1990s. Since then the budget has been fairly stable at a rather low level compared to the economic power of the United States and its claim to be the leading nation in this technology.
Negotiations on the Federal budget in the United States are a very complicated procedure involving of the Administration, the House of Representatives, and the Senate. It is remarkable that in many years the budget proposed by the Administration was indeed improved by Congress.
As has been mentioned before, funds from the DOE are allocated to grants for scientific research and to cost-shared contracts with industry, but in-house research at the national laboratories is also financed in this way. However, the maintenance of the national laboratories, which play a big role in the U.S. program, is not included in the PV budget.

Japan: The PV R&D budget in Japan is characterized by its very steady development over the years (Fig. 5.1b). The high figures in 1994 and 1995 in US dollars are due to the yen-to-dollar exchange rate. Budget development in yen looks quite smooth. The budget contains only financial grants outside the governmental laboratories, not in-house activities in the national laboratories.

Germany: Germany is the third partner with a long-term strong involvement in the financing of R&D, starting by the 1970s. The Federal budget had some ups and downs, but on average there was always a commitment to support new technology (Fig. 1c). The bump around 1990 to 1992 is additional money allocated to the "One-Thousand-Roofs-Project", introduced by the Federal Ministry of Research and Technology as a demonstration and field test program. This was the first "roofs" project, duplicated by other countries and by the new German "100 000 Roofs Program". But in contrast to the first program, which indeed had the character of a field test (monitoring of all systems), all other projects of this kind are instead market introduction programs, not R&D, and are therefore discussed in a later section.

It has to be mentioned that the diagram in Fig. 5.1c depicts only the Federal budget. In addition, a number of Federal States in Germany have their own R&D programs, with an appreciable amount of funding. They also pay for the maintenance of research institutes founded by these states.

Other European Countries: Many other European countries have PV R&D programs. Main players in Europe besides Germany have been France and Italy. Both spent between $100 and 200 million for R&D in the two decades from 1978 to 1998, and the historic development of their budgets are given in Tables 5.2a and 5.2b.

There has been additional money spent for research in the national laboratories. In France, CNRS and CEA currently contribute about 20 million FRF to R&D in the field of PV, the amount of basic funding for the Italian ENEA for PV is not exactly known, but may also be considerable because of the support of the institute Centro di Portici. Nearly all other European countries have supported PV research, but on a lower level than the main players. Their budgets are difficult to find, and no figures are presented here.

The European Union: The EU has a long tradition of funding R&D in the field of renewable energy. Many different programs have been conducted since the 1970s. For photovoltaics, between $10 and 20 million have been spent for R&D each year, mostly in cost-shared projects (50 %) with research institutes and industry. This is a relatively small budget compared to the big budgets of some countries, but it also helps to support some R&D in PV among the smaller partners in the EU. A detailed description of the activities of the EU follows in a later section.

Governmental Research Establishments The mechanism for distribution of funds for R&D is similar in the various countries. Most money is spent by grants to research institutions and by cost-shared research contracts with the industry. Some countries have national laboratories that are completely or partly devoted to R&D in renewable energy. Examples are listed below, and their research activities are described in a later section. In general, these institutions have appreciable basic funding from the government, which is not reflected in the PV budgets discussed in the previous section.

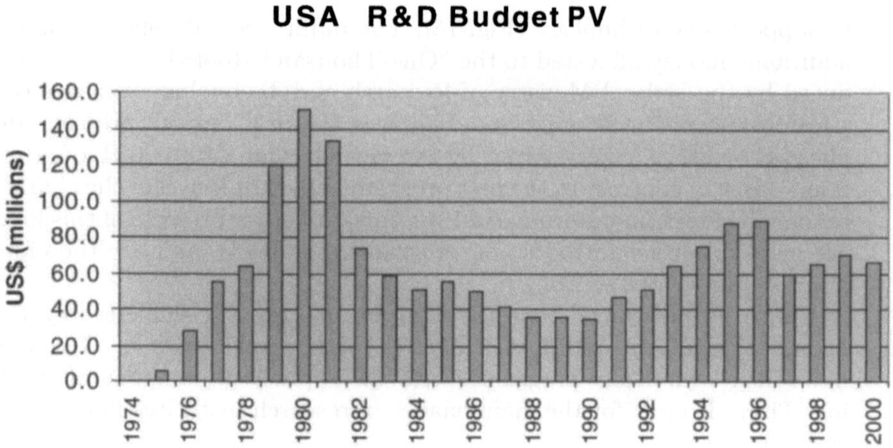

Fig. 5.1a. R&D budget of U.S.A

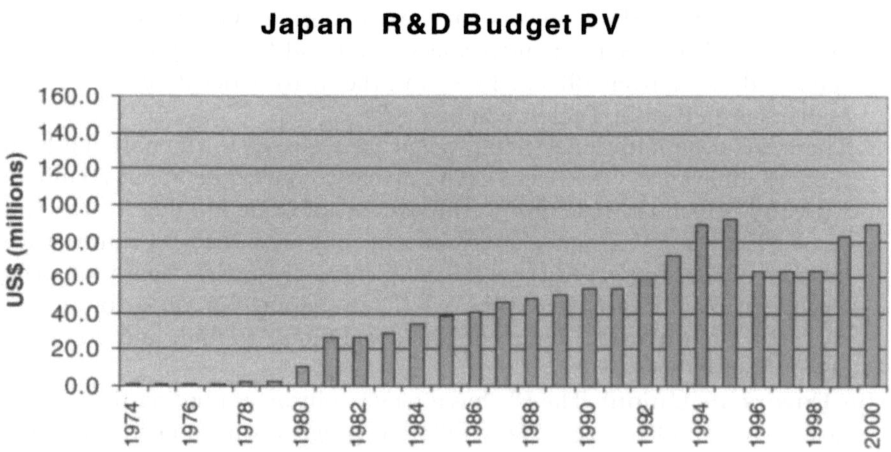

Fig. 5.1b. R&D budget of Japan

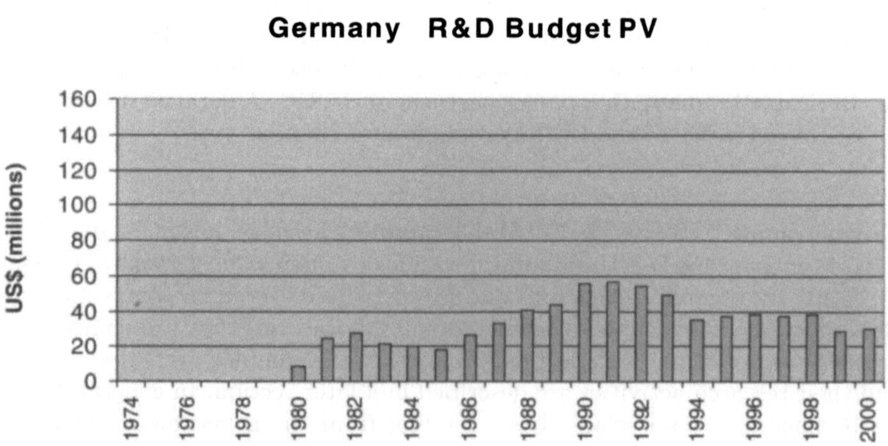

Fig. 5.1c. R&D budget of Germany

Table 5.2a. French R&D Budget handled by ADEME

Year	1981	1982	1983	1984	1985	1986	1987	1988	1989	1990
FRF (millions)	18	24	58	45	39	23.5	13	9.4	17.3	16.6
USD (millions)	3.3	3.7	7.6	5.2	4.3	3.4	2.2	1.6	2.7	2.1

Year	1991	1992	1993	1994	1995	1996	1997	1998	1999	2000
FRF (millions)	12.2	13.7	16.5	12.9	12.7	11.6	8	9	28	28
USD (millions)	2.2	2.6	2.9	2.3	2.3	2.2	1.4	1.5	4.5	4.5

Table 5.2b. Italian R&D Budget handled by ENEA

Year	by 1984	1985	1986	1987	1988	1989	1900
LIT (millions)		11 360	14 520	9 000	24 010	18 720	18 360
USD (millions)	36.83	5.97	7.63	6.95	18.48	13.66	15.35

year	1991	1992	1993	1994	1995	1996	1997	1998	1999	2000
LIT (millions)	17 120	20 200	22 700	36 141	32 460	39 600	35 000			
USD (millions)	13.86	16.40	14.47	22.44	22.23	21.05	23.27	20.19		

U.S.A.: In 1977, the Solar Energy Research Institute (SERI) was founded in Golden, Colorado. It expanded very rapidly. In the Photovoltaics Branch, the institute built up world-renowned expertise in materials research and in all thin-film technologies. In 1992, SERI was designated the National Renewable Energy Laboratory (NREL), underlining the importance of this institution. Besides NREL's own research activities, the institute is also engaged in the strategic planning of the PV program (National Photovoltaics Program Plan), in the administration of R&D contracts, and in the leadership of most subprograms.

Other national laboratories that have been or are active in the PV program include the Jet Propulsion Laboratory (JPL) in Pasadena, California, Sandia National Laboratories (SNL) in Albuquerque, New Mexico, and, to a minor extent, Brookhaven National Laboratory in Upton, New York. JPL has not further engaged since 1986.

Japan: The Electrotechnical Laboratory (ETL) in Tsukuba City is the main governmental research institute active in the field of photovoltaics. It reports directly to AIST.

Germany: There is no federal research institute devoted to renewable energy, but there is some activity on photovoltaics in two (former nuclear) research centres, the Forschungszentrum Jülich (FZJ) and the Hahn-Meitner-Institute (HMI) in Berlin. Both institutions are co-financed by the Federal Government and the Federal States. The largest PV research institute in Germany, the Fraunhofer Institute for Solar Energy Research (Fraunhofer ISE), is member of the Fraunhofer society (funded by the federal government and by the federal states: 90 % + 10 %). Three Federal States have founded research institutes for renewable energy on their own:

- ISFH in Hameln (by the State of Lower Saxonia)
- ISET in Kassel (by the State of Hessen)
- ZSW in Stuttgart (by the State of Baden-Württemberg)

Other European Countries: Many European countries have public institutes that are at least partly working in the field of photovoltaics. In France some Institutes of CNRS are engaged, in Belgium IMEC, in Italy the Centro di Portici, and in The Netherlands ECN, to mention just a few.

One rather important research establishment in Europe, connected to the European Commission, is the Joint Research Centre (JRC) in Ispra, Italy. Their PV department is responsible for tests and calibration, and JCR belongs to the few recognised institutions worldwide for the certification of modules.

5.3.3 The Organization of R&D Funding

Engaged Government Offices

U.S.A.: Since 1977 the whole program on renewable energy has been managed by the Department of Energy (DOE). Nowadays, DOE's Office of Energy Efficiency and Renewable Energy (EERE) is in charge of the PV program.

Japan: All energy issues are the responsibility of the Ministry of International Trade and Industry (MITI), and it handles the PV R&D program through its Agency of Industrial Science and Technology (AIST), whereas the PV dissemination program is managed by the Agency of Natural Resources and Energy (ANRE).

Germany: Until 1998, the energy research programs were the responsibility of the Federal Ministry of Education and Research (BMBF), formerly known as Federal Ministry of Research and Technology (BMFT). Since 1998, all energy research has been transferred to the Federal Ministry of Economics and Technology (BMWi).

European Union: Since 1. January 2000 the energy R&D programs of the EU have been organized by the Commission of the European Union through the newly formed Directorate General for Energy and Transport (DG-TREN). Formerly, the responsible directorates were DGXII and DGXVII.

Management Organisations In most cases the research programs are managed not by the Government offices but by an external organization.

U.S.A.: Nearly all management of the U.S. PV R&D program is conducted by the National Renewable Energy Laboratory (NREL). NREL assists in working out the strategies, they organize all research activities, and they also have responsibility for all contracts with industry.

Japan: The New Energy and Industrial Technology Development Organization (NEDO) was established under the MITI in 1980 as the principal semi-governmental organization for promoting R&D in energy and practical industrial technology. NEDO is responsible for the Japanese R&D program and also for the budget for research contracts, but the governmental research institutes like ETL and the Universities are financed directly by AIST. The Japanese PV Dissemination Program is handled by another organization, the New Energy Foundation (NEF).

Germany: The Project Management Organization Biology, Energy, and Environment (BEO) of the Research Centre Jülich (FZJ) organizes research for the Ministries BMBF and BMWi. The PV Branch of BEO handles all projects funded under the German PV R&D Program, but it also assists a number of international activities like the EU programs, and the IEA committees, and it manages bilateral cooperation with a number of countries, e.g. the ELDORADO Wind and Sun Programs.

France: The French Agency for the Environment and Energy Management (ADEME) handles the French PV R&D program. It guides projects with industry, the French utility EdF (Electricité de France), the Universities, and the research organization CNRS.

Italy: The Italian PV R&D Program is managed by the research organization ENEA.

5.3.4 Directions of Research

Process of Planning The planning of directions for R&D is handled differently by the governments in the various countries. It ranges from very strict long-term research plans with predefined goals (Japan) to very general guidelines in other countries (e.g. Germany). In the latter case research proposals from research institutes and from industry determine the direction, whereas in the case of binding research programs the administration finally determines the direction. Both models have their advantages and disadvantages. Long-term detailed planning can be very efficient if the right direction is chosen, whereas the models which prefer just general rules can react very smoothly to changes of condition.

Japan has very strict planning of their program. The main goal is to build a successful domestic PV industry which is able to fulfill the needs of the market development program. Very ambitious goals for technical achievements and for cost reductions in manufacturing are set by the funding agencies.

U.S.A.: The DOE has five-year planning periods, and the current National Photovoltaics Program Plan is valid from 2000 to 2004 [DOE 1]. The U.S. program is characterized by subprograms like PVMaT, PV-BONUS, PV Thin-Film Partnership, and others, which will be described in more detail later on. The general procedure is to have Calls for Proposals, followed by a negotiation phase and a decision by the authorities.

EU: The PV programs of the Commission of the European Union are part of the five-year RTD Framework Programs. These are described in more detail in a later section.

5.3.5 International Activities

Foreign Activities of Governments Since the beginning of PV development it has been realized that the countries of the sunbelt are ideally suited for the implementation of photovoltaic generators, and since the late 1970s all countries with a PV industry have promoted bilateral contacts with those areas. There were early cooperation agreements between the U.S.A. and Saudi Arabia, Mexico, Israel, Brazil, and other countries. Germany had a lot of activities in many countries, especially in Indonesia, the Philippines, Mexico, Marocco and Jordan. In the five years from 1991 to 1996, under the ELDORADO program, 0.5 MW_p of PV systems were installed in southern countries for test and demonstration purposes. France was pioneered the development and installation of solar home systems (SHS) in their Overseas Territories, mainly in French Polynesia. This is just a short selection of activities from the long list of programs and projects that could be cited over the last twenty years.

Although some of the programs look like market introduction efforts, they have their place in the R&D section. Installation of PV systems in hot and sometimes very humid climates is very demanding, and the early systems suffered from many difficulties and had a short life. Under the programs mentioned, the module companies learned how to fabricate modules that withstand extreme temperatures, heavy rains and storms, and other unfavorable conditions. Engineers had to learn how to build electric and electronic devices that can survive a tropical environment. All these projects form a long line of both disasters and successes, but they were needed to make photovoltaics a truely reliable technique.

Multinational Organisations

UNESCO For several reasons, the United Nations Educational, Scientific, and Cultural Organization has shown interest in solar energy, and especially photovoltaics, for a long time. Although UNESCO solar activities are typically development and aid projects, they also have some R&D aspects. UNESCO, in cooperation with the United Nations Development

Program (UNDP), assisted in the development of many types of standalone systems for use in remote areas, for use in medical stations, e.g. for vaccine refrigeration, and in schools and other educational establishments. The development of such systems at research institutions and in industry was financed by UNESCO or UNDP projects, as were tests at their final destinations.
Today, UNESCO activities have been incorporated into the World Solar Program 1996–2005 (WSP), and other organizations have been invited to join this ambitious effort to promote renewable energies in remote and undeveloped regions.

PEP The Photovoltaic Energy Project (PEP) was established by the G7 group of nations in order to exchange information on R&D activities in the field of photovoltaics and to initiate joint activities. PEP existed from 1981 to 1992, and the project had three main activities:
- Exchange of information on governmental activities like research programs, R&D budgets, main directions of research, and their results
- Two consecutive programs to develop common test and calibration procedures for solar cells. Three round-robin calibration campaigns were organized among all national calibration laboratories, and these experiments are the basis of today's agreement on solar cell measurements.
- Information exchange on health, safety and environment (HS&E) issues of PV technologies and products; relevant reports on studies were made available to the partner countries

IEA The International Energy Agency (IEA), an autonomous body within the Organization for Economic Cooperation and Development (OECD), has been engaged in solar energy issues since 1976. Today, most activity is in the IEA Photovoltaic Power Systems Program (PVPS), with the collaboration of 21 partner countries. This program is described in more detail in a later section.

5.4 Role of Industry

5.4.1 Engaged Industries

Companies engaged in photovoltaics come from several industrial branches: from the chemical industry, the semiconductor industry, the oil industry, and the utilities. Due to the large investment needed for the development of new technologies, PV is the domain of big concerns, rather than of smaller companies. The small and medium-size companies find their role in the field of systems integration, development of system components, and the development of consumer products.

Examples of the engaged big companies are listed below. The list contains also names of companies that played a big role in the development of photovoltaics but left the field later on. These are marked with an asterisk (∗).

Chemical Industry The chemical and metallurgical industry is mainly engaged in the development of silicon for solar cells, its crystallization, and its wafering. In the beginning, the big players were Wacker Chemie(*) in Germany, Union Carbide(*) and Hemlock(*) in the U.S., and Shin-Etsu Handotai(*) in Japan. They all went out and others came in, like Bayer AG in Germany, Kawasaki Steel, and Sumitomo Sitix in Japan. Elkem in Norway has been working in the field throughout.

Semiconductor Industry The solar cell is a semiconductor device, and one would expect strong engagement of the semiconductor companies. But this is not really true, and only in Japan are they dominat, with Kyocera, Sharp, Sanyo, Mitsubishi, and others. In Germany the first player was AEG-Telefunken(*), followed by Siemens, which is still one of the dominat companies in the field. Nothing has been left in the U.S. from the former engagement of semiconductor companies like Texas Instruments(*), RCA(*), and Westinghouse(*).

Oil Industry Oil companies played an important role from the beginning, and after some ups and downs, this role is now fairly stable, especially in the U.S. and in Europe. BP is among the market leaders, and Shell is coming up strongly. ARCO(*) and Mobil Oil(*) played a prominent role in the development of solar cell technologies, but neither companies exists any more.

Utilities Most activities of utilities are observed in Germany. The have both a negative and a positive influence. They discourage people and politicians with respect to the use of renewable energies; on the other hand, they invest actively in the photovoltaics industry. The German company ASE is fully owned by the utility RWE, and one half of Siemens Solar is owned by the utility Bayernwerk. In Italy, ENEL financed a lot of demonstration systems, whereas the French EdF is engaged in the development of systems for remote electrification.

5.4.2 Strategy of the Companies

Private companies always played an important role in research and development in the field of photovoltaics, and they had also a big influence on the research strategies of governments. The financial engagement of the private sector is certainly greater than the public money that was spent for R&D in the various countries. This was true for the early years of the new technology, when companies spent hundreds of millions of dollars for basic development, and it is true today when companies have to develop new manufacturing equipment.

Around 1976 it had become obvious that the crystalline silicon solar cell was the main option for photovoltaic electricity generation, and it also became clear that tremendous effort was needed to get the cost of the solar cell down to a level that allowed widespread application. This was the time to initiate big development programs for silicon feedstock (so-called "solar grade" silicon), for cheaper crystallization processes, and for cost-effective wafering. After ten years of effort most companies realized that the market for PV grew at a much slower rate than expected. On the other hand, cost reductions for materials can only be realized in large-scale production. Most developments in materials of that time came too early for the market, and most companies withdreaw from the field. Remarkable developments that have been left over are the crystallization of multicrystalline ingots (mc-Si), the wire saw for slicing wafers, and the EFG silicon ribbon material.

The early 1980s witnessed the beginning of the astonishing story of amorphous silicon. Although it was obvious from the beginning that the improvement of this material is extremely difficult, companies invested hundreds of millions of dollars in this technology. This was especially true for Japan, where most of the companies switched over to this new material, and the NEDO program was also focused on a-Si.

A few years later the view of the industry became more realistic. There was a renaissance of crystalline silicon technology, pushed by new efficiency records and innovative technologies and, on the other hand, other thin-film technologies came up. Since the early 1990s, a fairly well balanced development of all kinds of technologies has taken place.

The strategies of companies have certainly changed. A real market opportunity is now apparent, but the companies are more careful in their decisions about investments than they were twenty years ago. They know they have to develop the market in parallel with technical expansion. However, it is difficult to understand why most companies reduced their R&D efforts to a minimum even though it is obvious that the technologies are far from well developed.

5.4.3 Partnership Between State and Industry

Cooperative Planning of R&D Programs Today, photovoltaics are a global business, and all the big companies are engaged worldwide. On the other hand, governments want companies to be involved in their national programs. This is successful in some countries, but difficult in others.

Japan: Traditionally, the connection between industry and Government is very tight in Japan. It is not surprising that Japanese industry plays an important role not only in the planning procedure of the PV R&D program, but also in its management and control. High-ranking persons from industry are on important committees, but the main influence is exerted through industrial organizations like PVTEC and NEF, which are directly involved in the management of the program.

U.S.A.: The Department of Energy encourages industry to participate in Advisory Councils, and also in direct project planning. All larger sub programs like PVMaT, PV-BONUS, PVUSA, and the Thin-Film Partnership Programs have been worked out with the active collaboration of the related industry and utilities. In these partnerships, industry has free access to all research results of the National Laboratories.

In most other countries, there are no formal mechanisms for the cooperation of governmental offices with industry, but close links are often found: in France, the utility EdF is strongly linked to the rural electrification program of ADEME, and in Italy the big PV installations are mainly managed by the utility ENEL.

Market Opportunities Offered by Governments Besides today's real market development programs in Japan, Germany, and a few other countries, the R&D budgets offered from the beginning some special market in the field of demonstration and test installations. This is true for all countries and for a long time, and only some well-known larger projects are mentioned in the following. It should be pointed out that the market of demonstration systems is extremely important for the industry, not only because of its size. The main advantage is that it is the only possibility to test new products, to demonstrate the reliability of PV, and to be a showcase for future customers. The German One-Thousand-Roofs-Project is a very good example for the value of large-scale field tests. When it was finished, industry was really able to deliver reliable and well-designed grid-connected systems. This was not at all the case at the beginning of the program.

U.S.A.: "Block buys" of modules for use in demonstration projects were characteristic of the FSA Project. Nearly 2 MW_p of modules were bought from companies in the U.S. in five consecutive years from 1976 to 1980. Later on, the DOE co-funded the PVUSA Project with the installation of nearly one MW_p of test systems for grid interconnection. The following programs are mainly procurement programs, but in their initial phases many demonstration projects have been installed and financed by the R&D budget: Federal Buildings Program, National Park Program, Coast Guard, and Military Procurement Programs.

Japan: Under the Japanese R&D program, a large number of demonstration systems of all kinds have been installed. Two larger systems tests were the basis for further development, the Rocco Island Project with 200 grid-connected systems of about 2 kW_p each to study the behavior of a large number of PV systems in a grid, and the 1 MW_p Seijo Plant, a utility-type system. There is still a large testing and installation program funded under the AIST budget, mainly for installations on public buildings. The International Collaboration Project and also the multinational IEA Photovoltaic Power Systems Program comprise bilateral demonstration and test installations in a number of foreign countries.

Germany: The German PV Program funded many demonstration and test projects of all kinds in Germany, also outside Germany in the scope of bilateral projects with southern countries like Mexico, Jordan, and many others. The biggest effort at home was the "One-Thousand-Roofs-Project" funded by the Federal Government and the Federal States. This program led to the installation of more than 2000 systems on roofs of private houses, scattered around Germany, to serve as a large-scale field test.

France: For many years the French PV program was directed towards the development of systems for southern areas, mainly the former colonies. notable programs were the Sahel Program, which concentrated on water pumping in arid zones in northern Africa, and the DOM-TOM Program which concentrated on rural electrification in overseas territories mainly in the South Pacific. Only in the 1990s the French program start several projects in France, in cooperation with the French utility EdF. The FACE program pormotes demonstrations in the field of off-grid electrification, and the PHEBUS and HIP-HIP programs are engagements in grid-connected systems. The latter two programs are co-funded by ADEME, the European Commission, and EdF.

Italy: Characteristic of the Italian Program is the large number of utility-type demonstration systems. Most systems are co-funded by ENEA, the utility ENEL, and the European Commission. ENEL owns the biggest grid-connected system in the world, a 3 MW_p plant in Serre in southern Italy.

Other European countries: Many other countries have a lot of demonstration systems of different kinds, the highest density being in Switzerland and in The Netherlands.

5.5 Role of Research Establishments

5.5.1 Pioneers in the Development of New Technologies

From the beginning, research institutions have played a prominent role in the development of solar cells, PV systems, and systems components. In many fields they were the real driving forces and came up with many innovative concepts. Most research institutes have a rather intense collaboration with industry, and this helped to transfer the results of R&D in the manufacturing line.

This is not the place to appreciate all the activities in the research field in the different countries, but a few examples might illustrate the situation.

Silicon high-efficiency cells: The importance of efficiency improvement has always been pronounced at institutes like the University of New South Wales in Australia, the Fraunhofer Institute for Solar Energy Systems (ISE), the Georgia Institute of Technology in the U.S., the Instituto de Energia Solar in Spain, and many others.

Silicon manufacturing technologies: Strong institutes in this direction are IMEC in Belgium, ECN in the Netherlands, Sandia National Laboratory, and others.

Gallium arsenide and related III-V compounds: A prominent role is played by NREL in the U.S.

Amorphous silicon (a-Si): For many years Hamakawa's institute at the University of Osaka in Japan was the main driver in this technology. Today, one also has to mention activities in Europe, for example at the University of Neuchâtel in Switzerland and the Research Centre Jülich (FZJ) in Germany.

Compound thin-film cells: Most effort in thin-film research is certainly at the NREL in the U.S., both on CdTe and on the CIS family of materials. The big effort in Europe in the field of CIS technologies is headed by the Center for Solar Energy and Hydrogen Research (ZSW) in Stuttgart, Germany.

Concentrator systems: For many years the Sandia National Laboratory was very active in this field. In Europe, the main activities are at the Instituto de Energia Solar in Madrid.

This short list shows just a small section of the broad field of R&D, and many more examples would be worth mentioning. Especially in the field of systems and components development, the number of engaged institutions is very large.

5.5.2 Research Organizations

The collaboration of research institutions in the field of photovoltaics is well organized in most countries. We give examples are from the United States and Germany.

USA: First announced in November 1996, the National Center for Photovoltaics (NCPV) combines the effort of major research centers funded by the Federal Government. The headquarters is located at NREL, and NREL and with Sandia National Laboratory are the dominant partners. The other partners are Brookhaven National Laboratory, the Florida Solar Energy Center, the Southwest Technology Development Institute and DOE's Centers of Excellence in PV, the Georgia Institute of Technology, and the Institute of Energy Conversion at the University of Delaware.

Germany: In Germany, all major research institutions involved in solar energy research are cooperating in the Solar Energy Research Association (Forschungsverbund Sonnenenergie). The Members are

Hahn-Meitner Institute Berlin (HMI), Berlin
Fraunhofer Institute Solar Energy Systems (Fraunhofer ISE), Freiburg
Institute for Solar Energy Research (ISFH), Hameln
Research Centre Jülich (FZJ), Jülich
Institute for Solar Energy Techniques (ISET), Kassel
German Aerospace Center (DLR), Köln
Center for Solar Energy and Hydrogen Research (ZSW), Stuttgart
Plataforma Solar de Almeria (PSA), Almeria, Spain

The main activities of the association are the exchange of information, the organization of conferences, and public relations issues.

5.6 Selected Examples

5.6.1 The U.S. Federal PV Programme

The development of photovoltaics was totally dominated by the United States for the first twenty years following the invention of the solar cell in 1954. Military and civilian space programs were the driving forces, and by 1958 the first satellite with a PV array was launched (Vanguard I) immediately demonstrated the high performance and reliability of silicon solar cells, since Vanguard I transmitted signals to the ground for eight years. The development of space solar cells, with substantial funding from NASA and the military, gave the United States a very clear advantage for subsequent terrestrial applications.

As early as 1971, the U.S. federal government initiated the first terrestrial photovoltaic development program. The program was managed by the National Science Foundation (NSF), and later on by the Energy Research and Development Administration (ERDA).

In the 1970s, the energy policies of most developed nations were shaken by two oil crises, in 1973/74 and in 1979. They drastically demonstrated the strong dependence of most countries on energy imports. Both the U.S. federal government and the general public were certain that a big effort was needed to secure the nation's energy supply by the development of alternative energy sources. In addition, the idea of "clean energy" that could save the environment helped to initiate a really innovative program on renewable energy, and it was easy for the federal government to bring a rather big budget through Congress.

By 1973 the NSF had invited a number of well-recognised scientists, engineers, energy economists, and politicians to a meeting in New Jersey. At the famous Cherry Hill Conference it was clearly stated that photovoltaics is an energy option for terrestrial use in the U.S., and that the Government is encouraged to make efforts to develop this technology. The first appreciable funding was granted in 1976 ($28 million).

This positive development got its biggest push in 1978 when the Solar Photovoltaic Research, Demonstration, and Development Act passed the Congress and the 10-year Flat-Plate Solar Array Project (FSA) started.

Under the FSA program, JPL in Pasadena, California, managed most of the R&D funding of the federal government. The goal was to lay the groundwork for the actual commercial development of photovoltaics. JPL had very good experience in solar array development for space applications, and they were able to organize a concerted effort among research institutes and important industries in the terrestrial field.

The FSA Project has been subject to some criticism, because many of the early goals could not be reached. However, one has to keep in mind two important facts: at the beginning, the PV industry was really nonexistent or highly underdeveloped, as were the products, and secondly, in the early 1980s the program suffered severe budget cuts. Under these circumstances, the program was indeed a success. It analyzed the situation carefully, defined necessary research fields and started related projects, and it realized the necessity of quality tests for all components of systems and the need for demonstration. In fact, all subsequent projects in the U.S. and in other countries are more or less based on the knowledge gained during the FSA project.

The FSA was paralleled by five "block buys" of solar panels, as mentioned earlier. Severe tests were developed for the modules procured under the project, and the so-called "Block V Tests" formed the basis for all test procedures developed so far.

The FSA project concentrated on the development of crystalline silicon solar cells for flat plate applications. In parallel, there was a concentrator program led by Sandia National Laboratories and the start-up of a thin-film solar cell development at SERI. At that time, thin-film development was basic research, and this was left rather undisturbed during the severe budget cuts during the 1980s. As mentioned before, the policy of the Reagan administration excluded the funding of the commercialization of renewable energies and reduced the involvement of the Government in the funding of basic research. Even under these difficult conditions the U.S. PV Program remained well-organized and efficient, and the broad knowledge on the different technologies elaborated in research institutions and in industry is remarkable.

The years since 1991 have been characterised by a new national policy. The Energy Policy Act of 1992 and the Solar 2000 Strategy again enabled goverment support of the private sector to accelerate PV commercialization. Acronyms like PVMaT, PVUSA, and PV-BONUS stand for a successful program, in addition to the activities of the National Center for Photovoltaics (NCPV) and the Partnership Programs.

NCPV: All research institutions funded by DOE under the National Photovoltaics Program Plan are connected under the umbrella National Center for Photovoltaics. The partners of this association were described earlier. The primary goal of the NCPV is to help implement the Department of Energy's National Photovoltaics Program, which exists to support the U.S. PV industry in improving the cost effectiveness, performance, and reliability of its products.

Partnerships: The Thin-Film Partnership Program brings together the NREL Thin-Film Research Team with other R&D Partners and with industrial companies (Technology Partners) to find ways to bring thin-film technology to pilot production. The program is divided into four teams for the following areas: amorphous Silicon (a-Si), copper indium diselenide

(CIS), cadmium telluride (CdTe), and environment, safety, and health issues (ES&H). The program brings together the partners, manages the cost-shared projects, and organizes the exchange of information.

PVMaT: This project, which manages the cost-shared projects of DOE with industrial partners, is one of the most effective projects in photovoltaics. It is designed to help U.S. industry improve PV manufacturing processes and equipment, reduce manufacturing cost, and increase product performance and reliability. All major PV companies are partners in this project. The total contract sum since the 1992 starting round amounts to about \$140 million, with DOE funds of about \$80 millions. The DOE cost share accounts for about 58 %. Most innovative product improvements by the U.S. PV industry have been financed by the PVMaT project.

PVUSA: Photovoltaics for Utility Scale Applications (PVUSA) is a national cooperative research project established to demonstrate the potential use of photovoltaics in the utility environment. The project has been co-funded by DOE and a large number of utilities. This main activity was the installation and operation of a large test facility in Davis, California. Since 1992, a number of companies have delivered systems for long-term testing; the size of the systems ranges from $20\,kW_p$ to $400\,kW_p$. The funding of the project was about \$100 million, with at least \$30 million coming from DOE. The program reached all its goals, and after ten years of operation it might be closed in the near future.

PV-BONUS: Similar to the PVMaT Program, the program Building Opportunities in the U.S. handles cost-shared contracts between the DOE and industry. The aim of the program is to develop solutions for the integration of PV systems into buildings, in roofs, or in facades. There was a first round starting in 1992, with five partners from industry. The second round (BONUS II) started in 1998 with 18 contracts. As with most other programs, PV-BONUS is managed by NREL.

5.6.2 The Japanese Sunshine Project

As has been mentioned before, the Japanese PV Program started in 1974 with the Sunshine Project, based on a law that required action by the government in the development of new energy sources. The Moonlight Project was started in 1978 to develop energy conservation technologies. Finally, there was a third project, the Research and Development Project on Environmental Technology, which started in 1989. These three project merged into one big program, the New Sunshine Program, in 1993, in view of the close relationship between new energy development, energy conservation, and environmental measures. The development of solar energy technologies is only a small part of the overall concerted energy program, which covers the whole mixture of energy technologies available in Japan. The photovoltaic branch is the largest part of the solar energy division of the New Sunshine Project. Many organizations are working closely to make the program a success. All have been mentioned before and are only listed here to demonstrate the intensity of work in the program: MITI has the main responsibility for the whole program. The ministry acts through two agencies: the Agency for Industrial Science and Technology (AIST) is responsible for R&D, while the Agency for Natural Resources and Energy ANRE is responsible for market introduction. The management organization for AIST is NEDO, while ANRE handles the market introduction program through New Energy Foundation (NEF). Partners from industry are involved in all planning and management of the program through their organization PVTEC. The Japanese PV Program is characterized by thorough long-term planning with strong emphasis on good industrial strategy. The goal is "to develop innovative technology to create sustainable growth while solving energy and environmental issues." The result is clearly the very strong position of the Japanese PV industry, with its innovative technology and the largest production capacity in the year 2000.

5.6.3 The Activities of the European Union

The European Union (EU, formerly called European Community, EC) has a long-term commitment to the development of renewable energy sources. The first projects started by the 1970s. Objectives of the various programs always included the stimulation of research and technological development in the field of clean and efficient energy technologies, as well as the dissemination of those technologies. R&D Projects were funded under the so-called JOULE Programs, while the demonstration projects were collected under the THERMIE program. A third category involved the ALTENER projects, mainly studies of a strategic nature. All research within the responsibility of the Commission of the European Union is divided into five-year periods, the so-called Research and Technology Development (RTD) Framework Programs. Under the umbrella of these huge Framework Programs, a large number of sub-programs are carried out. Table 5.3 lists the most important programmes for PV R&D and demonstration.

The current 5th RTD Framework Program is subdivided into four Thematic programs, the relevant one being the Program Energy, Environment, and Sustainable Development. The Sub-program Energy contains Key Action 5: Cleaner Energy Systems, Including Renewables. The objectives are "to maintain and enhance the research potential of European laboratories, universities, and companies, and their ability to produce knowledge of the highest level and high-quality technologies." There are several calls for proposals, and projects financed are on a cost-shared basis (up to 50 % for research projects, up to 35 % for demonstration projects). Applications must be from at least two partners in at least two different EU partner countries. Although the total budgets of the EU PV programs have never been especially high, these programs had a big influence on the European PV community. They encouraged cooperation across the borders of the countries, and also of institutions. Research institutes learned to work together, and also with research

Table 5.3. History of EU Programmes in the Field of Photovoltaics.

Acronym	Framework Program	Years	Remarks
ENALT 1C		1979–1984	Exploitation of alternative energy sources
ENALT 2C		1983–1985	Exploitation of alternative energy sources
JOULE 1		1989–1992	Specific RTD in non-nuclear energy
JOULE 2	3rd FWP	1990–1994	Specific RTD in non-nuclear energy
THERMIE 1	3rd FWP	1990–1994	Promotion of energy (Demonstration)
RENA		1994	Preparatory action
ALTENER 1		1993–1997	Promotion of renewable energy sources
JOULE/ THERMIE	4th FWP	1994–1998	Specific RTD in non-nuclear energy
ALTENER 2		1998–2002	Promotion of renewable energy sources
ENERGY	5th FWP	1998–2002	Current Program

teams in industry. The programs also help to distribute knowledge around Europe, since partners can also join the projects from countries that cannot afford their own national research programs. The European programs focused mainly on solar cell technologies and PV systems development. A number of projects were notably successful and put European research teams in the first rank in several fields of solar cell research. To name just a few: MONOCHESS was a multi-year project on high-efficiency monocrystalline silicon solar cells, MULTICHESS was a similar project on multicrystalline silicon. EUROCIS and EUROCAD brought together nearly all groups in Europe working on CIS and CdTe thin-film solar cells. The budgets of the demonstration programs of the THERMIE series were very similar to those of the R&D programs. A large number of projects have been executed and helped to considerably improve systems technology. All types of systems from small standalone generators up to large utility-type plants have been installed. Well-known examples are the electrification of Greek islands, the electrification of hiker huts in the Alps, and numerous grid-connected medium-size generators in Italy. Co-funding from EU programs was also available for the two very big installations near Toledo, Spain, sized at 1 MW_p, and in Serre, Italy, at 3 MW_p. In summary, one can make the fair statement that the European RTD Programs were and are very beneficial in encouraging collaboration in Europe, and they were successful with respect to the improvement of the quality of research.

5.6.4 IEA Photovoltaic Power Systems Program

The International Energy Agency (IEA), founded in November 1974, is an autonomous body within the Organization for Economic Cooperation and

Development (OECD), which carries out a comprehensive program of energy cooperation among its member countries. Collaboration in research, development, and demonstration of new technologies has been an important part of the Agency's program. The IEA R&D activities are headed by the Committee on Research and Technology (CERT), and one of the four Working Parties, the Renewable Energy Working Party (REWP), is responsible for new energies and is charged with monitoring the various collaborative energy agreements. Soon after its establishment, the first engagement in solar energy started. Nineteen member states established the Solar Heating and Cooling Program in 1976. In the 1980s, TASK XVI of this program was the first IEA project on photovoltaics, devoted to Photovoltaics in Buildings. The next collaborative R&D Agreement, totally devoted to photovoltaics, was the IEA Photovoltaic Power Systems Program (PVPS). It started in 1993, and 21 member countries joined the program by the year 2000. The PVPS Agreement focuses on the pre-competitive and noncommercial aspects of the planning, design, construction, operational performance, and promotion of PV power systems. The mission of the IEA PVPS program is: "To enhance the international collaboration efforts through which photovoltaic solar energy becomes a significant renewable energy source in the near future."

The objectives of the PVPS Program are

- contribute to the cost reduction of their applications
- increase the awareness of their potential and value
- foster their market deployment by removing technical and non-technical barriers
- enhance technology cooperation with non-IEA countries

All kinds of power systems are under consideration: standalone systems to supply electric energy to isolated users and remote communities, both in rural areas of IEA member countries and in developing countries; Solar Home Systems, powering lights and television sets in about half a million rural households; and decentralized generators connected to the utility grid, either integrated into buildings and structures, or ground-based plants to provide grid support and peak power. Further cost reduction will encourage the ever larger deployment of these market segments, and in the long term PV will become a significant renewable energy option. The PVPS Program is divided into nine tasks, of which seven are operative in 2000. Task VI was completed in 1997, Task IV has not yet been initiated. The management of individual Tasks is the responsibility of Operating Agents. The Tasks are

Task I: Exchange and dissemination of information on photovoltaic power systems

Task II: Operational performance and design of photovoltaic power systems and subsystems

Task III: Use of photovoltaic power systems in standalone and island applications

Task IV: Design and grid interconnection of building-integrated and other dispersed photovoltaic systems
Task V: Design and operation of modular photovoltaic plants for large scale power generation, (concluded in 1997)
Task VI: Photovoltaic power systems in the built environment
Task VII: Very large scale photovoltaic power generation systems in remote areas
Task VIII: Technical cooperation for photovoltaic market deployment

In conclusion, the IEA Photovoltaic Power Systems Program is certainly very useful, since it forms an international platform for effective information exchange on technical, economic, industrial, and social issues.

5.7 Future

5.7.1 Further Need for R&D

The development of photovoltaics is certainly a success story, both from a technical and an economic point of view. The product has improved steadily, and the market for PV modules and systems expands at an appreciable rate. However, it has to be kept in mind that we are very far from our goal of making photovoltaics a viable energy option:

1. The technology of the solar cell is not a mature technology. There are still a large number of options, and the long-term winner is not obvious because all technologies have great potential for further improvements.
2. High-volume manufacturing techniques are not available, and much effort is needed to develop the appropriate equipment.
3. Effort is needed in many areas of systems technology. Examples are building integration and the improvement of energy management and storage for standalone systems.

5.7.2 Responsibility of the State for Further Progress

The gradual development of a self-supporting industry in the field of photovoltaics certainly does not mean that governments should pull out of this area and leave further research efforts totally to industry. The contrary is true. As in other modern technologies, it is in the primary interest of countries with their own PV industry to intensify research and education in that area. Funding of R&D must grow with the growing importance of a national industry.

A few points that should be considered in developing research plans are

1. Basic research in a broad sense is strongly required in photovoltaics, which is still in its infancy compared to other areas of semiconductor technology. The potential for technological progress is still very high.

2. Fast transfer of R&D efforts to the industrial scene is essential and should be encouraged as far as possible.
3. Education should be considered an important by product of R&D to guarantee that industry finds highly qualified scientists and engineers for their staff.
4. Expertise from industry should be included in the preparation of research plans, but instant need of industry must not be the only basis for long-term planning.

It is obvious and can be proven already by examples from the development in photovoltaics that well-developed R&D efforts are a good investment in a growing new technology and industry.

5.8 Further Reading

5.8.1 Information Sources

Books: The reader might wish to read a bit more on the history of PV development, and for the first steps a few articles/books are mentioned here:

William Wallace
Government Terrestrial Acceleration Programs
in: Solar Cells and Their Applications, Ed. Larry D. Partain, John Wiley & Sons, Inc., New York, 1995

Alain L. Fahrenbruch and Richard H. Bube
Fundamentals of Solar Cells
Academic Press, New York

Paul D. Maycock and Edward N. Stirewalt
A Guide to the Photovoltaic Revolution
Rodale Press, Emmaus, PA

John Perlin
From Space to Earth: The Story of Solar Electricity
Aatec Publications, Ann Arbor, Michigan, 1999

Conferences: There are three series of conferences to be mentioned which build the forum for the presentation of most results in R&D in photovoltaics. Each of these series is held in intervals of 18 months. The proceedings of these meetings are essential sources of information.

1. IEEE Photovoltaic Specialists Conferences in the U.S.A. The 28th Conference was held in September 2000 in Anchorage, Alaska.
2. The European Photovoltaic Solar Energy Conference and Exhibition. The 16th Conference of this series was held in May 2000 in Glasgow, Scotland.
3. The Asia/Pacific PV Science and Engineering Conferences. The 11th Conference was held in October 1999 in Sapporo, Japan.

Two times until now the three conferences have been combined in a World Conference on Photovoltaic Solar Energy Conversion. The first happened in 1994 in Hawaii, the second was held in 1998 in Vienna, Austria.

6 Strategies in Photovoltaic Research and Development – Market Introduction Programs

F. Staiß and A. Räuber

6.1 Introduction

The main barriers to market introduction are still the high electricity cost of grid-connected systems and the lack of appropriate financing schemes for standalone systems in developing countries. To reduce these barriers programs have been launched, apart from R&D projects, in many countries.

In industrialized countries, financial promotion programs concentrate on grid-connected systems. They can be divided into incentives per electricity output and incentives per capacity. In other words, the two major strategies to give financial support are to subsidize electricity production (rate-based incentives) or to influence investment cost (rebate and loan programs). In the past, both types have been applied and several historical milestones have stimulated PV market introduction.

In the developing world, many countries also have programs to foster renewable energy technologies and PV in particular. In most cases, programs are implemented in bilateral or multilateral corporation with donors. In general, strategies clearly focus on small-scale rural electrification, since PV is the supply alternative that should be favored in many cases, not only for economic but also for ecological and social reasons.

6.2 Market Introduction Efforts in Industrialized Countries

Most industrialized countries are characterized by the existence of a relatively dense electricity grid, leaving only very few areas with dwellings far from the common grid, for example in the mountains, in arid regions (e.g. in the southwestern USA), or on islands (e.g. in the Mediterranean Sea). Except for these areas, introduction of photovoltaic systems is mainly discussed as a supplementary electricity source in a grid-connected arrangement. Here, privately owned smaller and decentralized systems are favored over larger centralized systems in the property of utilities or other companies.

As has been pointed out in the previous chapter, the very first governmental effort to regulate the so-called "third party" generation of electricity with connection to the common grid seems to be the PURPA law. The "Public Utility Regulatory Policies Act" was introduced in 1978 to regulate the relationship between utilities and their customers who produce part of their electricity consumption themselves. The utility had to allow the connection of other generators to the grid. Moreover, they had to provide backup power, and to buy excess power produced by the "co-generators" at reasonable rates. Normally, utilities used the "avoided cost scheme" to buy electricity from third parties, which is in general much lower than the selling tariff of the utility. The most important measures to foster the introduction of PV electricity generation are summarized in Table 6.1.

Table 6.1. Measures of governments to foster the market introduction of PV

Regulatory Measures

- Obligation of the grid company to allow co-generation
- Obligation of the grid company to allow the feed-in of electricity from independent power sources into the grid
- Regulations regarding the safety of PV systems
- Regulations regarding the quality of the electricity fed to the grid (frequency, phase, harmonic content)
- Adaptation of building codes to the installation of PV systems on buildings
- Quality assurance of PV components (certification)

Financial Measures

- Investments
  Subsidies as percentage of systems cost
  as a fixed amount per system
  Low-interest loans ("soft loans")
- Taxation
  Tax credits (reduction of the tax debt by a certain amount)
  Depreciation allowance for private taxpayers to apply higher depreciation rates on PV plants than on other commercial investments
  Low import tax (in countries without their own PV production, e.g. in developing countries)
- Feed-in tariffs
  - Obligation to buy fed-in electricity at least on an avoided cost basis
  - Enhanced feed-in tariffs (also called "rate-based incentives")
    - fixed tariffs (by order of government)
    - cost-related tariffs
  - Net metering
    - equal tariffs for buying and selling electricity realized by two-way meters (show only the difference between bought and fed-in electricity
- Combination of instruments

By 1982 the first metering program (i.e. the electricity meter runs backwards when solar electricity is fed into the grid) was introduced in the USA. While it addressed only a very limited number of consumers, the first wide-ranging field test program was Germany's 1 000-Roofs Program that started in 1989 and led to 2 250 systems with an overall capacity of more than 4 MW_p until 1993.

In the early 1990s the Sacramento Municipal Utility District (SMUD) in California introduced green pricing schemes that were also been applied in other countries later on. At about the same time the first rate-based incentives were launched in Switzerland and Germany. Also in the early 1990s Switzerland's ENERGIE-2000 Program attracted attention by aiming at an installed PV capacity of 50 MW_p (i.e. about 5 W_p per capita) by the end of 2000. In 1994 the biggest public program for PV in the first half of the 1990s, the Japanese 70 000 rooftop program, was launched; several thousands of systems have been installed by now.

In the middle of the 1990s solar stock exchange and bidding models were established in the cities of Zurich (1995) and Berlin (1997). Also in 1995, so-called full cost rate compensation became popular when it was introduced in the German city of Aachen. In this model the full cost (at that time almost $1/kWh) of PV electricity fed into the public grid is compensated. By now this communal regulation has been taken over by more than 30 other municipalities in Germany, and also by a few in Austria and Switzerland (where the model was first developed). In 1997, in Germany alone, this instrument led to an additional capacity of several megawatts. In 1998 public purpose programs in California were introduced, focusing on the promotion of renewables in a liberalized electricity market. By the end of 1998, the Spanish government had enacted a law according to which compensation rates amount to $0.36/kWh for solar systems with an installed capacity of less than 5 kW, and $0.18/kWh for systems between 5 kW and 50 MW. In the late 1990s some new big rebate programs also attracted attention: President Clinton's initiative to equip 1 million roofs in the USA with solar systems (also including solar water heaters), the Italian 10 000 Rooftop-Program, and Germany's 100 000-Roofs Program for PV, which was launched in 1999.

In Europe the following main types of incentives are practiced today: on the public level, rebate/loan programs and regulated rates, and on the private level, facultative "green" pricing. Rebate programs have been practiced or will be practiced e.g. under the various rooftop programs in Austria (200 kW), Italy, Germany, and Switzerland. Here, a certain percentage of up to 50–70 % of total investment is subsidized by authorities and/or utilities. Instead of rebates soft loans are also offered, e.g. in Germany under the 100 000-Roofs Program or under the programs of some German Federal States.

The rebate programs mainly have in view private households, since their willingness to pay for PV electricity is in general much greater than that of commercial investors. On the other hand, some dozens of larger plants have been erected since the early 1980s. They have been funded either by national/European programs (e.g. THERMIE) and/or utilities/private companies, and have largely served as demonstration projects. Practical experiences, tests, and demonstrations of new technical developments have been in the foreground.

Meanwhile, premium rates for electricity from renewables have been legally established in most European countries, but they show a wide variety. Although they have helped other renewable energy technologies, mainly wind power, energy rates have obviously not been sufficient to cover additional costs of PV systems, except under the new Spanish rule and local full cost rates. But a powerful new incentive was established in Germany in spring 2000: in the process of further developing the feed-in law, which was established in 1991, the new Renewable Energy Law subsidizes PV electricity production by 0.99 DM/kWh (about \$0.5/kWh). This decision will create a new quality of financial promotion which may push PV to a similar market development as has been the case with wind energy in the last few years.

6.3 Programs in Developing Countries

Many countries in the developing world also have programs to foster renewable energy technologies, and PV in particular. In most cases, programs are implemented in bilateral or multilateral cooperation with donors and clearly focus on small-scale rural electrification, since PV is the supply alternative that should be favored in many cases, not only for economic but also for ecological and social reasons. Programs address a variety of technical, financial and institutional problems that have to be overcome. However, a successful PV household electrification program must overcome the high initial cost of solar systems: the purchase price of a solar home system (SHS), which can range from \$100 (10 W_p China) to \$1 400 (53 W_p Kenya) for many low- to middle-income households represents almost one year's income. Thus there is a need to develop and implement appropriate financing schemes (including grants and subsidies) for this target group, which often has little or no access to credit.

A number of rules should govern the implementation policy:

1. Establish a sustainable infrastructure: successful PV programs need financial sustainability, effective management, and support services. This can only be achieved if they can be operated as a business by the private sector, i.e. in the long term, prices should allow full cost recovery. Moreover, qualified managers and technicians are needed which often requires extensive training. Appropriate tools and transportation, as well as locally available spare parts have to be provided.
2. Provide quality products: an important issue for success is the quality of products. They have to be well designed, properly assembled and installed, and have to meet consumers' needs and capacity to pay.

3. Ensure appropriate support from governments and donors: since it is difficult to build up a small PV industry by the private sector alone, assistance from government and donors is required. Support can include conducting least-cost rural energy planning that includes PV system options, making investment capital available for PV programs, encouraging the commercial banking sector and financing agencies to finance systems on reasonable terms, supporting promotional campaigns among rural households, removing regulatory barriers that limit competition among energy service providers, and offering training and technical assistance to establish retail and service networks.

Obviously, international programs have created a lot of confidence in PV during the last few years. Although this development has not been sustainable, big progress has been made in creating national PV markets. India may serve as an example, presently being the most important PV market in the developing world. In India, more than 450 000 PV systems have been installed, accounting for 40 MW_p. At the moment, 75 organizations are involved in manufacturing PV equipment, and a small domestic PV industry has been established, which has the capacity to produce 25 MW_p PV cells. A key role is played by the World Bank Group, which has been involved in many rural PV projects throughout the world.

6.4 Special Examples

Japan and Germany are the countries with the largest and most efficient market introduction programs. These are described in more detail in this chapter.

6.4.1 Germany

There exist indeed a large variety of market introduction schemes by

- the federal government
- the federal states
- local authorities (cities, local utilities, etc.)

6.4.2 Federal Government

Early actions of the Federal Ministry of Research and Technology include many demonstration projects and especially the 1 000-Roofs Program, which was intended to serve as a large field test of grid-connected PV systems installed on rooftops of one-family residences. Due to the size of the program (more than 2000 systems have been installed) it indeed had an impact on the German PV market and triggered the growth of a number of small engineering and installation companies.

The 1 000-Roofs Program was co-financed by the federal government and the federal states, the largest part remaining with the federal government. Investors received subsidies of up to 70 % of the investment.

The Government also introduced, and with better long-term prospects, the Electricity Feed Law (Stromeinspeisungsgesetz), with an obligation for the utilities

- to allow the connection of PV systems to the grid
- to buy the fed-in electricity at a fixed buy-back rate

This law applied to all renewable energies at various buy-back tariffs. For wind energy and PV the rate was fixed at 90 % of the average selling tariff of utilities, resulting in a rate of around 0.17 DM/kWh. This rate made wind energy nearly economical, but for the expensive PV technology it was just a first step.

The utilities could recover their enhanced costs by slight enhancements of their selling tariffs, but nevertheless they fought vigorously against the law. Regional differences, especially for wind energy generation indeed caused some unbalance.

When the 1 000-Roofs Program came to an end in 1993, the PV market in Germany broke down for a certain time until other financing schemes came up, especially on a local basis (see below).

Appreciable efforts started again in 1998 when the Government announced the 100 000-Roofs Program which is a low-interest loan program. Investors can receive a nearly zero-interest loan for 10 years from the governmental bank KfW (Kreditanstalt für Wiederaufbau) for the installation of PV systems.

The second action with even more impact was the Renewable Energy Law (EEG Erneuerbare Energien Gesetz) introduced in spring of 2000. This law regulates the feed-in of electricity from renewable sources and replaced the old Stromeinspeisungsgesetz.

This new law is very favorable for PV, since the buy-back rate of 0.99 DM/kWh is fairly high and guaranteed over 20 years. In the years 2002 and 2003 this rate will be reduced for new plants in increments of 5 %.

The EEG has some important features:

- The enhanced rate is paid by the owner of the grid. These entities are separate from the power producers according to the deregulation practice in Germany.
- Not only private customers get the high rate but also commercial companies and big power producers.
- The enhanced cost for renewable energies is distributed equally to all customers in Germany to avoid regional differences.

The combination of the 100 000-Roofs Program with the Renewable Energy Law has already had a tremendous impact on the German PV market.

Thus, one can expect that the political target to install 350 MW_p of PV capacity in grid-connected applications will be reached by the year 2003. Although the law limits the obligation to pay the above mentioned feed-in rates for new plants to this volume, it is quite probable that there will be appropriate regulation for the years beyond to continue further market expansion.

6.4.3 Federal States

All 16 German federal states are engaged in the promotion of renewable energy, but with strongly differing intensity. Their involvement started with the cooperation with the Federal Government in the scope of the 1 000-Roofs Program. Some of the states had or have their own funding programs for investments, others offer low-interest loans by their state banks.

In total, a market volume of at least 24 MW_p initiated by programs of the states up to 1999, the most active state being Nordrhein-Westfalen. A few years ago, some of the states were acting not directly but through their state utilities. Good examples are the State of Bavaria through the utility Bayernwerk, and Berlin through BEWAG. These utilities introduced subsidy programs, sometimes under a special label like "PV for Schools" or other special applications. Others, like the State of Hamburg through their utility HEW, introduced the scheme of cost-oriented buy-back rates, i.e. higher buy-back rates than the Electricity Feed Law (before 2000) but no full cost rates.

Due to the privatization of the utilities, and due to the advent of efficient programs from the federal government, most federal states reduced their market introduction programs considerably, and only a few states are still engaged in the year 2000.

6.4.4 Local Authorities

In Germany, a large variety of local efforts exists with respect to the introduction of renewable energies. Many cities and even villages have their local programs, in many cases handled by the municipal utilities (Stadtwerke). Most of the programs find their roots in pressure exerted by citizens on their communities, asking for some local actions for the environment and the climate.

Actions in the cities range from subsidies to private customers for investment in PV systems, subsidies for the installation of systems on public buildings, to very generous rules on the buy-back rates for fed-in electricity. A well-known example of the latter case is the so-called Aachen model. The basic feature of this model is that fed-in electricity is bought by the utility at a rate that makes the investment in a PV system fully economical. Full-cost feed-in tariffs were around 2 DM per kWh in the early 1990s, and they are constant over the whole life of the system. Due to the price reduction of PV systems, the full-cost rate for new systems has been adjusted every year to a reduced level.

A number of other German cities adopted the full cost Aachen model, others used some enhanced or cost oriented feed-in rates between the legal lower limit set by the former Electricity Feed Law (0.17 DM/kWh) and the highest rates set by the Aachen model. These enhanced rate models were the main reason for the growing German market after the end of the 1 000-Roofs Program.

Local activities played a very important part in the market development of photovoltaics in Germany, and activities encouraged by the municipal authorities have been among the most successful market players, although their importance naturally declined with the introduction of the 100 000-Roofs Program and the Renewable Energy Law. On the other hand, local activities shifted to a certain extent to individual private activities, since after the liberalization of the German electricity market in 1998, a number of "green electricity tariffs" were offered either by the traditional electric utilities or by new suppliers who specialize in such products. In general, private customers pay an extra charge on their standard tariff of the order of 0.05 to 0.10 DM/kWh, of which the bigger part is invested in new renewable generation plants. In a number of cases it is foreseen - according to the regulations of some certification institutions - to reach a PV share of 1 % of the green electricity generation mix. In total, the importance of this kind of private engagement for PV is still very little compared to the big public programs since only a few tens thousands of German households (in almost 40 million) have ordered green tariffs. But this might change in the long term.

6.4.5 Japan

As mentioned in the previous chapter, the Japanese PV program started in 1974 with the Sunshine Project, based on a law that required action by the Government in the development of new energy sources. The Japanese PV program is characterized by thorough long-term planning with strong emphasis on good industrial strategy. The goal is "to develop innovative technology to create sustainable growth while solving energy and environmental issues."

The New Sunshine Project, initiated in 1993, defined a new energy policy, which included goals for the market development of renewable energy. The basic guidelines for new energy introduction set the year 2000 objective for PV system installations at 400 MW_p. This target will not have been fully reached by the end of fiscal year 2000 because of limitations on the supply side, but surprisingly strong development did take place.

The Japanese program introduced by MITI has in fact two parts. The major program is called "Introduction and Promotion of Residential PV systems," the other two very similar programs are called "PV Field Test Program for Industrial Use," and "PV Field Test Program for Public Facilities."

The Residential Program is handled by the New Energy Foundation (NEF), the Field Test Programs by the New Energy and Industrial Technology Development Organization (NEDO). All programs are based on subsidies for investment, by providing a certain percentage of the investment, or by a certain fixed amount of subsidy per installed power. The latter model was used during the last years in the residential program, and the amount of subsidy was reduced every year as the price of the systems came down. In 2000 the subsidy amounted to 180 000 Yen/kW installed power.

The budgets of the three programs are shown in Table 6.2.

The residential program supports the installation of grid-connected PV systems on private houses. It belongs to the series of "roof projects" initiated by the German 1 000-Roofs Program. In Japan, besides the PV industry, the housing industry plays a role in the program, especially in financing schemes for the systems.

The program has thus far been very successful. There was tremendous interest in the program among house owners, and in most years the program was overbooked. The mean size of the systems installed was around 3.5 kW_p.

The status of the program near the end of the year 2000 is shown in Table 6.3.

In summary, the Japanese program is indeed a success: about 90 billion Yen have been spent to install nearly 100 000 systems with a total installed power of nearly 300 MW_p. During this program the Japanese PV industry built up appreciable production capacity, and they are able to serve a big market. It is, however, true that this market is still not self-supporting, and it seems essential that at this stage the promotion policy continues, for a further period of time. Mass production techniques are still to be developed, and further cost reductions have to be realized.

6.5 Conclusion

Market introduction programs are an important part of the efforts of governments to improve the role of renewable energy in the primary energy mix of their countries. Main motivations are the introduction of "clean" energies to protect the environment and the climate, and also the reduction of the country's dependence on energy imports.

A number of measures have been used so far to facilitate the installation of PV systems and to stimulate the market.

Regulatory measures comprise technical codes for safe electrical connection, building codes for installation of the PV generator on roofs or other places, and rules for the connection of these systems to the public grid.

A number of models for financial subsidies exist to facilitate the relatively high investments still needed for PV systems. Investment subsidies and soft loans are direct measures to ease the financial burden for the private investor, but special taxation measures (tax credits, enhanced depreciation) can also be very effective. Only very recently, several governments introduced the model of enhanced feed-in tariffs for electricity generated by renewable energy sources like wind, PV, and others. This seems to be a very effective way to

Table 6.2. Budget of MITI for three PV introduction programs (in billion Yen)

	FY 1994	FY 1995	FY 1996	FY 1997	FY 1998	FY 1999	FY 2000	FY 2001 (request)	total
Residential Program	1.97	3.31	4.06	11.11	14.70	16.04	14.50	23.60	89.00
Field Test Industrial Use					2.40	2.41	4.00	2.04	10.85
Field Test Public Facilities	1.03	1.70	1.90	1.35	0.17	0.11	0.10	0.07	6.43

The Japanese Fiscal Year (FY) starts in April and ends in March of the following year

Table 6.3. Status of the Japanese PV Residential Program.

		FY 1994	FY 1995	FY 1996	FY 1997	FY 1998	FY 1999	FY 2000	FY 2001 (request)	total
Budget	(Billion Yen)	1.97	3.31	4.06	11.11	14.70	16.04	14.50	23.60	89.00
Systems	(Number)	539	1 065	1 986	5 654	6 252	17 396	*23 016*	*37 460*	93 468
Power Installed	(MW)	1.9	3.9	7.5	19.5	24.1	60.9	80.6	131.1	272.6

Figures in italics are estimates.

Table 6.4. Web sites for topical information about current governmental programs

Acronym	Organisation	Country	Web Address
EU	EU R&D Information Service CORDIS	Europe	www.cordis.lu
ADEME	Agency for Environment and Energy Management	France	www.ademe.fr
BMWi	Federal Ministry of Economics and Technology	Germany	www.bmwi.de
BEO	Project Management Organisation Biology, Energy, Environment	Germany	www.fz-juelich.de/beo
ENEA	Ente per le Nuove Tecnologie, l'Energia e l'Ambiente	Italy	www.enea.it
MITI	Ministry of International Trade and Industry	Japan	www.miti.go.jp
AIST	Agency for Industrial Science and Technology	Japan	www.aist.go.jp
NEF	New Energy Foundation	Japan	www.or.jp
NEDO	New Energy and Industrial Technology Development Organisation	Japan	www.nedo.go.jp
EREN	DOE PV Programme	U.S.A.	www.eren.doe.gov/pv/
DOE	Department of Energy	U.S.A.	www.doe.gov
NCPV	National Center of Photovoltaics	U.S.A.	www.nrel.gov/ncpv/
NREL	National Renewable Energy Laboratory	U.S.A.	www.nrel.gov
IEA	IEA Photovoltaic Power Systems Progamme	World	www.iea-pvps.org

stimulate the introduction of these energy sources, since the additional cost is equally distributed among all customers.

Japan and Germany are the front-runners in the market introduction of PV systems, but a number of mainly European countries are following their lead. The impact of the current programs on the growth of the market and the production capacity of the industry is remarkably strong.

References

[Raynolds 1954] D.C. Raynolds, G. Leics, L.L. Antes and R.E. Marburger: Photovoltaic effect in cadmium sulfide. Phys. Rev. 96(1954).

[Chapin 1954] D.M. Chapin, C.S. Fuller and G.L. Pearson: A new silicon p-n-junction photocell for converting solar radiation into electric power. J. Appl. Phys. 25(1954)676.

[Jenny 1956] D.A. Jenny, J.J. Loferski and P. Rappaport: Phys. Rev. 101(1956)1208.

[Bonnet 1972] D. Bonnet and H. Rabinhorst: Proc. 9th IEEE PVSC, 1972, p.129.

[Wagner 1974] S. Wagner, J.L. Shay, P. Migliorato and H.M. Kasper: Appl. Phys. Lett. 25(1974)434.

[Kazmerski 1975] L.L. Kazmerski, M.S. Ayyagari and G.A. Sanborn: J. Appl. Phys. 46(1975)4865.

[Carlson 1976] D.E. Carlson and C.R. Wronski: Appl. Phys. Lett. 28(1976)671.

[Staebler 1977] D.L. Staebler and C.R. Wronski: Appl. Phys. Lett. 31(1977)292.

[Wallace 1995] William Wallace: Government Terrestrial Acceleration Programs. in: Solar Cells and Their Applications, Ed. Larry D. Partain, John Wiley & Sons, Inc., New York, 1995.

[DOE 2000] The National 2000-2004 Photovoltaics Program Plan Report DOE/GO-10099-940, January 2000-07-27.

Only a few references are given at this place since most information can easily be traced in the sources presented before (see also Table 6.4).

A Acronyms and Abbreviations

ADEME	Agence de l'Environnement et de la Maîtrise de l'Energie (France)
AIST	Agency of Industrial Science and Technology (MITI, Japan)
AP	Acidification potential
AR	Antireflection
a-Si	Amorphous silicon
BEO	Project Management Organization Biology, Energy, and Environment (Germany)
BMBF	Federal Ministry of Education and Research (Germany)
BMFT	Federal Ministry of Research and Technology (Germany; now BMBF)
BMWi	Federal Ministry of Economics and Technology (Germany)
BOS	Balance of system
CdTe	Cadmium telluride
CED	Cumulative energy demand
CERT	Committee on Research and Technology of the IEA
CIGS	Copper indium gallium di-selenide
CIS	Copper indium di-selenide
CNRS	Centre National de la Recherche Scientifique (France)
CORECT	Committee on Renewable Energy Commerce and Trade
CRIEPI	Central Research Institute of Electric Power Industry (Japan)
c-Si	Crystalline silicon
Cz-Si	Czochralski silicon (monocrystalline)
DLR	Deutsches Zentrum für Luft- und Raumfahrt
DOE	US Department of Energy
ECN	The Netherlands Energy Research Foundation
EdF	Electricité de France
EEG	Renewable Energy Law (Germany)

EERE	Office of Energy Efficiency and Renewable Energy (DOE, USA)
EFG	Edge-defined film-fed growth
eg-Si	Electronic grade silicon
ENEA	Ente per le Nuove Tecnologie, l'Energia e l'Ambiente (Italy)
EPBT	Energy payback time
ERDA	Energy Research and Development Administration (USA, 1970s)
EREN	DOE PV Program, `www.eren.doe.gov/pv/`
ETL	Electrotechnical Laboratory (Japan)
EU	European Union
EVA	Ethylene vinyl acetate
EYR	Energy yield ratio
FSA	US Flat-plate Solar Array Project
FZJ	Research Centre Jülich (Germany)
FZ-Si	Float-zone silicon (monocrystalline)
GWP	Global warming potential
HMI	Hahn-Meitner Institute (Berlin, Germany)
HDR	Hot-dry-rock technology
HS&E	Health, Safety, and Environment
IEA	International Energy Agency
IMEC	Inter-University Microelectronics Centre (Belgium)
IOA	Input/Output analysis
I_{sc}	Short circuit current
ISE	Fraunhofer Institut für Solare Energiesysteme (Freiburg, Germany)
IPCC	International Panel on Climatic Change
JRC	Joint Research Center of the European Union in Ispra, Italy
kW_p	One thousand W_p
LCA	Life cycle assessment
LCI	Life cycle inventory
Mc-Si	Multicrystalline silicon
mg-Si	Metallurgical grade silicon
MITI	Ministry of International Trade and Industry (Japan)
m-Si	Monocrystalline silicon
MW_p	One million W_p
NCPV	National Center for Photovoltaics (USA)
NEDO	New Energy & Industrial Technology Development Organization (Japan)

NEF	New Energy Foundation (Japan)
NELT	Net energy lifetime
NMHC	Non-methane hydrocarbon
NREL	National Renewable Energy Laboratory (USA)
NSF	National Science Foundation
OECD	Organization for Economic Cooperation and Development
PECVD	Plasma enhanced chemical vapor deposition
PERL	Passivated emitter rear locally diffused
P_{mpp}	Power at maximum power point
p-Si	Polycrystalline silicon
PV	Photovoltaic
PVMaT	Photovoltaic Manufacturing Technology
RES	Renewable energies
R&D	Research and development
RD&D	Research, development and demonstration
REWP	Renewable Energy Working Party of the IEA
RIGES	Renewable Intensive Global Energy Scenario
SEE	Solar Energy Economy Scenario
SEIA	Solar Industry Association
SERI	Solar Energy Research Institute (USA, became NREL in 1992)
SHS	Solar Home System
SNL	Sandia National Laboratory (USA)
sog-Si	Solar-grade silicon
StrEG	Stromeinspeisungsgesetz (German electricity feed-in law)
TCO	Transparent conductive oxide
TCS	Trichlorosilane
UNESCO	United Nations Educational, Scientific, and Cultural Organization
UNDP	United Nations Development Program
UPVG	Utility Photovoltaics Group (USA.)
V_{oc}	Open circuit voltage
W_p	Electrical power of solar cells, PV modules, or systems under standard testing conditions
WEC	World energy conferences 1995 and 1998
ZSW	Center for Solar Energy Research and Hydrogen Technology (Germany)

B Participants of the Reisenburg Workshop

Peter Adelmann
Phocos AG
89171 Illerkirchberg, Germany

Dr. Hubert Aulich
PV-Silicon GmbH
99099 Erfurt, Germany

Ulrich Beyer
RWE-Power AG
45118 Essen, Germany

Dr. Harry Blask
Bundesministerium für Wirtschaft
und Technologie,
Referat Energieforschung
Erneuerbare Energien
53123 Bonn, Germany

Dr. Dieter Bonnet
Antec Solar GmbH
99334 Rudisleben, Germany

Prof. Dr. Achim Bubenzer
Fachhochschule Ulm
89075 Ulm, Germany

Günther Cramer
SMA-Regelsysteme GmbH
34266 Niesetal, Germany

Bernhard Dimmler
Würth Solar GmbH & Co. KG
71672 Marbach, Germany

Dr. Bernhard Dreher
Umweltbundesamt Abt. III 2.2
14191 Berlin, Germany

Dr. Manfred Fischedick
Wuppertal Institut für Klima,
Umwelt und Energie
42004 Wuppertal, Germany

Dr. Karin Jahn
Bremer Energie Institut
28359 Bremen, Germany

Prof. Dr. Jürgen Garche
Zentrum für Sonnenenergie-
und Wasserstoff-Forschung (ZSW)
Baden-Württemberg
89081 Ulm, Germany

Dr. Paul-Georg Guthermut
53604 Bad Honnef, Germany

MinRat Ulf Hecksteden
Ministerium für Wissenschaft und
Forschung Baden-Würtemberg
70029 Stuttgart, Germany

Dr. Georg Hörning
Akademie für
Technikfolgenabschätzung
70565 Stuttgart, Germany

Prof. Dr. Werner Kleinkauf
Universität Gesamthochschule Kassel
34109 Kassel, Germany

Dr. Frithjof Kublik
Deutsche Shell AG
22297 Hamburg, Germany

Prof. Dr. Joachim Luther
Fraunhofer Institut für Solare Energiesysteme (ISE)
79100 Freiburg, Germany

Jochen Möller
Universität Oldenburg
26111 Oldenburg, Germany

Dr. Ing. Martin Müller
EnBW
76180 Karlsruhe, Germany

Dr. Joachim Nitsch
Deutsches Zentrum für Luft- und Raumfahrt e. V. (DLR)
70569 Stuttgart, Germany

Gernot J. Oswald
Siemens Solar GmbH
80807 München, Germany

Martin Pehnt
Institut für Energie- und Umweltforschung
69120 Heidelberg, Germany

Dr. Armin Räuber
PSEmbH
79114 Freiburg, Germany

Dr. Horst Schade
Angewandte Solarenergie – ASE GmbH
Division Phototronics
85640 Putzbrunn, Germany

Evi Pschorr-Schoberer
Ludwig-Bölkow-Systemtechnik GmbH
85521 Ottobrunn, Germany

Dr. Hans-Werner Schock
Universität Stuttgart
Institut für physikalische Elektronik
70569 Stuttgart, Germany

Dr. Thomas Schott
Zentrum für Sonnenenergie- und Wasserstoff-Forschung (ZSW)
Baden-Württemberg
70565 Stuttgart, Germany

Dr. Gerd Stadermann
Forschunsverbund Sonnenenergie
12489 Berlin, Germany

Dr. Frithjof Staiß
Zentrum für Sonnenenergie- und Wasserstoff-Forschung (ZSW)
Baden-Württemberg
70565 Stuttgart, Germany

MinDirig Dr. Armin Tschermak von Seysenegg
Wirtschaftsministerium
Baden-Württemberg
70029 Stuttgart, Germany

Prof. Dr. Fritz Vahrenholt
Deutsche Shell AG
22297 Hamburg, Germany

Prof. Dr. Ulrich Wagner
Forschungsstelle für Energiewirtschaft
80995 München, Germany

Walter Weber
E. ON Energie AG
80333 München, Germany

Prof. Dr. Peter Woditsch
Deutsche Solar GmbH
47829 Krefeld, Germany

Dr. H. J. Ziesing
Deutsches Institut für
Wirtschaftsforschung
14195 Berlin, Germany

Index

MIX
Papier aus verantwortungsvollen Quellen
Paper from responsible sources
FSC® C105338

If you have any concerns about our products,
you can contact us on
ProductSafety@springernature.com

In case Publisher is established outside the EU,
the EU authorized representative is:
Springer Nature Customer Service Center GmbH
Europaplatz 3, 69115 Heidelberg, Germany

Printed by Libri Plureos GmbH
in Hamburg, Germany